Solution
of

Ordinary
Differential
Equations

by
Continuous
Groups

Solution
of
Ordinary Differential Equations

by
Continuous Groups

George Emanuel

CRC Press
Taylor & Francis Group
Boca Raton London New York

CRC Press is an imprint of the
Taylor & Francis Group, an **informa** business

A CHAPMAN & HALL BOOK

Published 2018 by CRC Press
Taylor & Francis Group
6000 Broken Sound Parkway NW, Suite 300
Boca Raton, FL 33487-2742

First issued in paperback 2019

ISBN-13: 978-1-58488-243-5 (hbk)
ISBN-13: 978-0-367-39787-6 (pbk)

Library of Congress Cataloging-in-Publication Data

Emanuel, George.
    Solution of ordinary differential equations by continuous groups /
George Emanuel.
      p. cm.
    Includes bibliographical references and index.
    ISBN 1-58488-24-3 (alk. paper)
    1. Differential equations--Numerical solutions. 2. Continuous
groups. I. Title.
QA372 .E43 2000
515'.352—dc21

00-050748

Library of Congress Card Number 00-050748

Visit the Taylor & Francis Web site at
http://www.taylorandfrancis.com

and the CRC Press Web site at
http://www.crcpress.com

Preface

The theory of continuous groups is employed to solve ordinary differential equations (ODEs). The presentation is self-contained and presumes for background only a rudimentary exposure to ordinary differential equations. A prior knowledge of group theory, algebraic concepts, or the method of characteristics is not required; when needed, these topics are developed in the text. The book is thus intended for upper division or graduate students in applied mathematics, engineering, or the sciences. As a book on applied mathematics, it may be useful as a monograph. The author is an engineer, not a mathematician, and is unaware of any university course that treats the subject matter of this book. The principal audience is probably readers who are self-instructing themselves. If you are in this category, be sure to examine the worked examples at the end of the chapters in Part II.

A number of topics, such as the theory of a complete or Jacobian system of partial differential equations (PDEs) and the theory of contact transformations, are occasionally discussed in Lie group monographs. These topics, however, are extraneous to our goals and are not covered. Background material of a relatively advanced nature (for an engineer) has been trimmed to what is essential. This enables us to focus on the central issue: solving ODEs by group methods.

The text is replete with many fully worked examples, and homework problems are provided at the end of the chapters. Answers to selected problems are contained in Appendix E. Mathematical derivations and proofs are provided but are deemphasized relative to problem-solving. Every effort, however, has been made to retain the conceptual basis of continuous groups and to relate the theory to problems of interest in engineering and the sciences.

Over a century ago, Sophus Lie used continuous groups to develop a systematic transformation theory for finding separation of variable solutions for ODEs, or similarity solutions for PDEs. The theory has not found widespread use for solving ODEs, where typically the handbook of solutions by Polyanin and Zaitsev (1995) would be consulted. (See the bibliography in Appendix A.) This monograph hopes to rectify this oversight by providing an application-oriented version of the theory.

As an engineer, my efforts on this monograph began in 1986, but were quite sporadic. In the course of this lengthy process, a number of new techniques have been developed, which are published here for the first time. Probably the most useful of these is the enlargement procedure in Part II, and all of the material in the several compendium sections in Part II. A new way of organizing tables is also introduced. The purpose of these contributions is to significantly increase the number of ODEs amenable to a group-theory solution. In addition, by imposing an organizational scheme, reminiscent of what is used for integral tables, the method should be easier to implement.

Most books on Lie groups are quite abstract, and if they deal with differential equations focus on PDEs. A good introduction to the subject is contained in Ibragim (1994), which is nevertheless at a more advanced level than this monograph. Moreover, this book does not discuss the substitution principle, the enlargement procedure, or the compendium process. These topics are inextricably linked with the practical solution of ODEs by group methods.

As is usual with a book that began as lecture notes, a number of former students contributed comments and corrections. Particularly helpful was Dr. J. Rodriquez. I also wish to thank Profs. K. A. Grasse and F. Civan, of the University of Oklahoma, for their comments and encouragement. I am indebted to Shirley Doran, of Texas Christian University, for her expert typing of the manuscript. Any further comments or corrections, especially with regard to the tables, would be most welcome.

Contents

Part I

BACKGROUND

Chapter 1

Introduction

There is no single systematic way of analytically solving a nonlinear ordinary differential equation (ODE). Rather, there is a collection of tricks or techniques, some of which are:

(1) change of variables,

(2) separation of variables,

(3) exactness condition,

(4) integrating factor,

(5) variation of parameters, etc.

Some of these items are related, e.g., 3 and 4. The most common and powerful technique is the first one in which the dependent and independent variables are transformed. Our hope, often unrealized, is that after the transformation a first integration by separation of variables, item 2 in the list, is possible.

The idea of separation of variable is quite simple. Suppose we have a first-order ODE in its most general form

$$f(x, y, dy/dx) = 0, \tag{1.1}$$

where f is an arbitrary function of its three arguments. If this equation can be written as

$$M(y)dy = N(x)dx,$$

where M is a function only of y and N is a function only of x, then there is a solution

$$\int M \, dy = \int N \, dx + a, \tag{1.2}$$

where a is a constant of integration. This separation of variables solution is in terms of two quadratures, in which neither, one, or both may be integrable in terms of standard functions. Even if one or both of the quadratures cannot be analytically performed, we still view Equation (1.2) as the general solution of Equation (1.1). We thus accept one or more quadratures as part of a solution,

3

where the quadratures can be numerically evaluated, approximated analytically, or written in terms of known functions. Although the discussion has focused on a first-order ODE, the technique of separation of variables is applicable to ODEs of any order.

In the later part of the 19th century, a Norwegian mathematician, Sophus Lie, developed an ingenious transformation theory for finding separation of variable solutions for ODEs. His approach utilizes the properties of continuous groups. This book is an exposition of this theory. Our presentation is slightly based on Cohen (1911) and Dickson (1923). Additional reading material is provided in the bibliography, Appendix A. (Some of these references also discuss the application of group theory to partial differential equations, which we do not consider.)

For many decades, Lie algebras have been a major research area of mathematics. As a consequence, the theory is highly abstract and obtuse (since I don't comprehend it), involving far more mathematical apparatus than is warranted for the solution of ODEs. Undoubtedly, this is one of the reasons for its failure to significantly impact any of the engineering disciplines. We shall introduce only the most essential elements of group theory. In line with this orientation, many illustrative examples are provided, often to the exclusion of formal mathematical proofs. This approach is most apparent when algebraic concepts, which are needed for some derivations, are not covered. Many of the omitted proofs are unnecessary for engineering or scientific purposes, since their validity is (heuristically) evident, especially when the theory is applied to specific problems. We hope that this approach will make up in clarity and readability what it may lack in formal rigor.

An example of Lie theory would be to find the general solution of the nonlinear equation

$$\frac{d^2y}{dx^2} + ke^y = 0, \tag{1.3}$$

where k is a constant. As it stands, this ODE cannot be integrated even once by separation of variables. Without guesswork, however, the theory provides the substitution

$$X = y, \quad Y = y', \tag{1.4}$$

where a prime hereafter indicates differentiation; hence, y' is a shorthand for dy/dx. The second derivative transforms as

$$\frac{d^2y}{dx^2} = \frac{dy'}{dx} = \frac{dY}{dy}\frac{dy}{dx} = Y\frac{dY}{dX}$$

with the result

$$Y\frac{dY}{dX} + ke^X = 0.$$

This relation is separable and integrates to

$$Y^2 + 2ke^X = a, \tag{1.5}$$

where again a is a constant of integration.

The above equation is a first integral of Equation (1.3). We solve it for Y and use Equations (1.4) with the positive square root result

$$\frac{dy}{dx} = (a - 2ke^y)^{1/2}.$$

This equation is also separable; its solution is

$$x = b + \int \frac{dy}{(a - 2ke^y)^{1/2}},$$

where b is a second integration constant. While many quadratures cannot be performed in terms of elementary functions, here the substitution

$$z^2 = a - 2ke^y$$

$$2z\,dz = -2ke^y\,dy = (z^2 - a)\,dy$$

allows for evaluation as

$$x = b + 2 \int \frac{dz}{z^2 - a} = b + \frac{1}{a^{1/2}} \ln \left(\frac{z - a^{1/2}}{z + a^{1/2}} \right)$$

$$= b + a^{-1/2} \ln \left[\frac{1 - \left(1 - \dfrac{2k}{a}e^y\right)^{1/2}}{1 + \left(1 - \dfrac{2k}{a}e^y\right)^{1/2}} \right].$$

Upon inversion, we obtain

$$y = \ln \left\{ \frac{a}{2k} \operatorname{sech}^2 \left[\frac{1}{2} a^{1/2}(x - b) \right] \right\}$$

for the final form for the solution of Equation (1.3). The integration constant a appears in a complicated manner because the ODE is nonlinear. This is in sharp contrast to the linear appearance of all integration constants in the solution of any linear differential equation. Another point worth noting is that the solution is singular when $k \to 0$, whereas Equation (1.3) has the solution

$$y = c_1 x + c_2,$$

where the c_i are integration constants.

The key step in the above example is the use of the Transformation (1.4), which is provided by group theory and enabled us to perform a first integration. The theory is relatively easy to use when it is known to apply. This last proviso is crucial since it may not apply. It is applicable whenever an ODE is invariant under (or with respect to) a continuous transformation group, such as a rotation about the origin. A differential equation, however, may not possess group invariance. This limitation is not surprising, since for nonlinear ODEs there is no systematic analytic approach that guarantees a solution in terms of quadratures.

The theory is applicable to both linear and nonlinear ODEs, although nonlinear equations are clearly of greater interest. As just noted, the ODE must be invariant under, at least, one group for application of the theory. An extensive series of tables are provided in conjunction with the chapters in Part II to help satisfy this criterion. Because the theory is usually not practical for third- and higher-order ODEs, the presentation is largely, although not completely, limited to first- and second-order ODEs.

Every attempt has been made to introduce a clear and consistent notation. The numerous worked examples are designed to clarify the discussion, illustrate how the theory is applied, and to provide continuity between chapters. This latter point is evident in our frequent use of the rotation group, which appears in many chapters. For the reader's convenience, this material is summarized in Appendix B. Similarly, for the reader's convenience, Appendix C contains key equations and the equation number where they first appear.

Despite the emphasis on applications, Part I is devoted to basic background material, such as the group concept, infinitesimal transformations, invariance, and the method of characteristics. By the way, this latter topic will play a central role in the subsequent analysis.

Part II is devoted to ODEs, culminating in Chapter 7, which treats second-order equations invariant under a two-parameter group.

Part III constitutes the appendices. These include the bibliography (Appendix A), rotation group (Appendix B), and a summary of key equations (Appendix C). In addition, all tables have been relegated to Appendix D, since these are frequently needed for solving problems, and it is therefore convenient to have them in one location. The last appendix contains the solutions to some of the assigned problems.

Chapter 2

Continuous One-Parameter Groups-I

This chapter develops basic results for a continuous group that is generated by a free parameter, hereafter denoted as α. Each element of the group therefore corresponds to a specific value of this parameter. The group is *continuous* because α can vary continuously over the real numbers. Our discussion begins by first defining an arbitrary group, after which we focus on continuous groups.

2.1 Group Concept

A *group* is a finite or infinite collection of elements together with an operation called *group multiplication*. By definition, a group G has four properties:

(1) The group is *closed* under the operation $\circ$. This means that $T_i \circ T_j$ is an element in G for any elements T_i and T_j in G.

(2) There exists in G a unique *identity* I, such that for any T_i in G, $T_i \circ I = I \circ T_i = T_i$.

(3) Each T_i in G has a unique *inverse* in G denoted by T_i^{-1}, such that
$$T_i^{-1} \circ T_i = T_i \circ T_i^{-1} = I$$

(4) The *associativity* property
$$T_i \circ (T_j \circ T_k) = (T_i \circ T_j) \circ T_k$$
holds.

The positive and negative integers, including zero, constitute a group where the operation is addition. The identity element is zero and the inverse of n is $-n$. A simple geometric example of a group involves the rotation in its own

7

plane of a circle about it center. For instance, the collection of rotations through multiples of $2\pi/n$ radians constitutes a finite group with n distinct elements, where n is an integer. Here, the elements of the group are the discrete points on a circle, while the operation is the $2\pi/n$ rotation.

The groups of interest to us are the *continuous transformation groups*, where an individual group element is a transformation. The operation in this case is called *composition*, which means the application of successive transformations. As an example, consider translation parallel to the y-axis

$$x_1 = x, \quad y_1 = y + \alpha \tag{2.1}$$

in the x, y plane, where the real number α is the *group parameter*. With translation as the group element, we examine the above criteria, starting with:

$$T_i : x_1 = x, \quad y_1 = y + \alpha_1,$$

(1)

$$T_j : x_2 = x, \quad y_2 = y + \alpha_2.$$

Then $T_i \circ T_j$ is the translation

$$x_3 = x, \quad y_3 = y + \alpha_3, \quad \alpha_3 = \alpha_1 + \alpha_2,$$

which is in the group.

(2) The identity element is $\alpha = 0$.

(3) The inverse is $(-\alpha)$. Thus, every translation has a unique inverse translation.

(4) The sequential transformations

$$x_1 = x, \quad y_1 = y + \alpha_1$$
$$x_2 = x_1, \quad y_2 = y_1 + \alpha_2$$
$$x_3 = x_2, \quad y_3 = y_2 + \alpha_3$$

or

$$x_3 = x, \quad y_3 = [(y + \alpha_1) + \alpha_2] + \alpha_3$$

provides the same result as

$$x_2 = x_1, \quad y_2 = y_1 + \alpha_2$$
$$x_1 = x, \quad y_1 = y + \alpha_1$$
$$x_3 = x_2, \quad y_3 = y_2 + \alpha_3$$
$$x_3 = x, \quad y_3 = (y + \alpha_1) + \alpha_2 + \alpha_3.$$

Hence, associativity is established as a result of the associativity of addition for real numbers. Consequently, translation parallel to the y-axis constitutes a continuous group, as does translation parallel to any straight line in the x, y plane.

A second example of a continuous group is a rotation about the origin

$$x_1 = x \cos \alpha - y \sin \alpha,$$
$$y_1 = x \sin \alpha + y \cos \alpha, \tag{2.2}$$

where the identity transformation is given by $\alpha = 0$, and the inverse transformation has $-\alpha$. The transformation merely rotates point x, y counterclockwise in a circular arc about the origin by an angle α. We shall often refer back to this transformation when illustrating the subsequent theory.

Both examples are *point transformations*, which can be written as

$$x_1 = \phi(x, y, \alpha), \quad y_1 = \psi(x, y, \alpha). \tag{2.3}$$

This transformation constitutes an element of a group. For brevity, these equations are referred to as a group. Hereafter, we shall use $\alpha = 0$ to denote the *identity transformation*

$$x_1 = \phi(x, y, 0) = x, \quad y_1 = \psi(x, y, 0) = y \tag{2.4}$$

of the group. There is no loss of generality in this choice, since α can always be redefined to yield these equations. Similarly, the *inverse transformation* is written as

$$x = \phi(x_1, y_1, \overline{\alpha}), \quad y = \psi(x_1, y_1, \overline{\alpha}). \tag{2.5}$$

When Transformation (2.3) satisfies group properties, it is referred to as a *one-parameter continuous group in two variables*. The subsequent development for ordinary differential equations will focus on this type of transformation.

Not all continuous one-parameter transformations have the group property. For instance,

$$x_1 = \frac{\alpha}{x}, \quad y_1 = y$$

has the inverse, $\overline{\alpha} = \alpha$, but no identity transformation, nor does it satisfy closure. An especially noteworthy aspect of groups is that linearity is not required or even intrinsic to the concept.

For the continuous, one-parameter transformations of interest, we really need to verify only three of the four properties. For these transformation groups, associativity is guaranteed, since the composition of transformations is an associative operation. Of course, there are other types of groups where associativity represents an independent requirement.

Groups, in general, are often not commutative, as in matrix multiplication. Here, we do not require commutivity, although many of the groups are commutative. (A commutative group is usually referred to as Abelian.) This is apparent in the so-called *law of composition*

$$\alpha_3 = F(\alpha_1, \alpha_2), \tag{2.6}$$

where α_1 and α_2 stem from two successive transformations

$$x_1 = \phi(x, y, \alpha_1), \quad y_1 = \psi(x, y, \alpha_1),$$
$$x_2 = \phi(x_1, y_1, \alpha_2), \quad y_2 = \psi(x_1, y_1, \alpha_2).$$

By closure, we have

$$x_2 = \phi(x, y, \alpha_3), \quad y_2 = \psi(x, y, \alpha_3)$$

for some α_3, where α_3 is a function of α_1 and α_2. For the preceding explicit examples, Equation (2.6) becomes

$$\alpha_3 = \alpha_1 + \alpha_2.$$

When

$$\alpha_3 = F(\alpha_1, \alpha_2) = F(\alpha_2, \alpha_1)$$

the group is commutative. It is also worth noting that the existence of a unique inverse means the *transformation is one-to-one*.

Another important facet of group transformations is that the group property is invariant under a change of coordinates that is single valued. Suppose we have the group given by Transformation (2.3) and a one-to-one coordinate change

$$u = f(x, y), \quad v = g(x, y).$$

Because it is one-to-one, its inverse can be written as

$$x = r(u, v), \quad y = s(u, v).$$

Then, the transformed group is

$$u_1 = f(x_1, y_1) = f(\phi(x, y, \alpha), \psi(x, y, \alpha))$$
$$= f(\phi(r(u, v), s(u, v), \alpha), \psi(r(u, v), s(u, v), \alpha))$$

with a similar equation for v_1. We thus obtain

$$u_1 = \tilde{\phi}(u, v, \alpha), \quad v_1 = \tilde{\psi}(u, v, \alpha),$$

which represents the transformed version of Equations (2.3). These equations satisfy group properties. For instance, closure is readily established. This invariance property is of crucial importance in the later analysis. *It means that group methods for solving ODEs are independent of the choice for the coordinate system.*

2.2   Infinitesimal Transformation

A one-parameter group is a global transformation in that x_1 and y_1 may differ considerably from x and y, respectively. Of greater utility is the corresponding infinitesimal transformation. In this type of transformation, a small change in α results in a small change in x_1 and y_1, as shown in Figure 2.1. Observe that the x, y point is kept fixed. We thus have

$$x_1 + \delta x_1 = \phi(x, y, \alpha + \delta\alpha),$$
$$y_1 + \delta y_1 = \psi(x, y, \alpha + \delta\alpha).$$

Expand ϕ in a Taylor series, to obtain to first order

$$x_1 + \delta x_1 = \phi(x, y, \alpha) + \frac{\partial\phi}{\partial\alpha}\delta\alpha + \cdots$$
$$= x_1 + \frac{\partial\phi}{\partial\alpha}\delta\alpha + \cdots$$
$$\delta x_1 = \frac{\partial\phi}{\partial\alpha}\delta\alpha,$$

and similarly for y_1

$$\delta y_1 = \frac{\partial\psi}{\partial\alpha}\delta\alpha.$$

The partial derivatives are evaluated for the identity transformation, when $\alpha = 0$. Hereafter, this is denoted by an o subscript. Hence, the *infinitesimal transformation* is

$$\delta x = \left.\frac{\partial\phi}{\partial\alpha}\right|_o \delta\alpha = \xi(x, y)\delta\alpha,$$
$$\delta y = \left.\frac{\partial\psi}{\partial\alpha}\right|_o \delta\alpha = \eta(x, y)\delta\alpha, \tag{2.7}$$

where ξ and η will prove of great importance. (They are referred to as *infinitesimal elements*.) Their definition is

$$\xi(x, y) = \left.\frac{\partial\phi}{\partial\alpha}\right|_o, \quad \eta(x, y) = \left.\frac{\partial\psi}{\partial\alpha}\right|_o. \tag{2.8}$$

In general, both ξ and η are not identically zero. They are both zero only when ϕ and ψ are independent of α, and Transformation (2.3) no longer represents a one-parameter group.

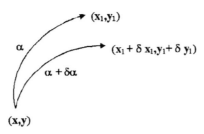

Figure 2.1. Schematic of an infinitesimal transformation.

For translation parallel to the y-axis, Equations (2.1), we obtain

$$\xi = \left.\frac{\partial \phi}{\partial \alpha}\right|_o = 0, \quad \eta = \left.\frac{\partial \psi}{\partial \alpha}\right|_o = 1. \tag{2.9}$$

For rotation, we have

$$\phi = x \cos \alpha - y \sin \alpha,$$

$$\xi = \left.\frac{\partial \phi}{\partial \alpha}\right|_o = (-x \sin \alpha - y \cos \alpha)_o = -y, \tag{2.10a}$$

and

$$\psi = x \sin \alpha + y \cos \alpha,$$

$$\eta = \left.\frac{\partial \psi}{\partial \alpha}\right|_o = (x \cos \alpha - y \sin \alpha)_o = x. \tag{2.10b}$$

Let us examine how an arbitrary function $f(x, y)$ changes as a result of an infinitesimal transformation. We thus have

$$f(x + \delta x, y + \delta y) = f(x, y) + \frac{\partial f}{\partial x}\delta x + \frac{\partial f}{\partial y}\delta y + \cdots .$$

But δx and δy are given by Equations (2.7), so that

$$\delta f = f(x + \delta x, y + \delta y) - f(x, y) = \left(\xi\frac{\partial f}{\partial x} + \eta\frac{\partial f}{\partial y}\right)\delta \alpha + \cdots ,$$

or to first order

$$\delta f = (\xi f_x + \eta f_y)\delta \alpha, \tag{2.11}$$

where we denote $\partial f/\partial x$ as f_x and $\partial f/\partial y$ as f_y. It is convenient to call the differential operator, U, defined by

$$Uf = \xi f_x + \eta f_y \tag{2.12}$$

the *symbol* of the infinitesimal transformation. In fact, Uf represents the infinitesimal transformation in that knowing Uf yields the transformation and vice versa. Since the global, or finite, transformation is related in a one-to-one manner with its infinitesimal transformation (as will be evident shortly), the symbol also represents the global transformation, even though it is independent of α.

In the previous examples, Uf is

$$Uf = f_y \tag{2.13}$$

$$Uf = -yf_x + xf_y \tag{2.14}$$

for translation parallel to the y-axis and rotation, respectively. A brief list of several elementary groups and their symbols is provided in Table 2.1, see Appendix D. The geometrical group interpretation is also given, although this aspect is of little importance for the subsequent analysis.

For later reference, it is worth noting the general result

$$Ux = \xi, \quad Uy = \eta. \tag{2.15}$$

These relations are utilized in the following section.

2.3 Global Group Equations

Suppose the symbol is known and we wish to find the global transformation. In terms of solving differential equations, it is the symbol and not the global equations that is important. Nevertheless, the results of this section, such as the Lie series, are essential in the later analysis. Furthermore, a knowledge of the global transformation may be useful in physically interpreting the solution of the related differential equation. For use in the later analysis, two methods are provided for finding the global equations. The Lie series formula is first developed, since it is the basis of the first method.

Consider a Taylor series expansion about $\alpha = 0$ for some function f

$$f_1 = f(x_1, y_1)$$

$$f_1 = f_{1,o} + \alpha \frac{df_1}{d\alpha}\Big|_o + \frac{\alpha^2}{2!} \frac{d^2 f_1}{d\alpha^2}\Big|_o + \cdots .$$

As before, an o subscript denotes evaluation at $\alpha = 0$. Since $\alpha = 0$ denotes the identity transformation, we have

$$f_{1,o} = f(x_1, y_1)\big|_o = f(x, y)$$

$$\frac{df_1}{d\alpha}\Big|_o = \left(\frac{\partial x_1}{\partial \alpha} \frac{\partial f_1}{\partial x_1} + \frac{\partial y_1}{\partial \alpha} \frac{\partial f_1}{\partial y_1} \right)_o = \xi f_x + \eta f_y = Uf$$

$$\frac{d^2 f_1}{d\alpha^2}\Big|_o = \left(\frac{\partial x_1}{\partial \alpha} \frac{\partial}{\partial x_1} + \frac{\partial y_1}{\partial \alpha} \frac{\partial}{\partial y_1} \right)^2 f_1\Big|_o = U^2 f$$

$$\vdots$$

We have used, when $\alpha = 0$,

$$f_1 = f, \quad x_1 = x, \ldots ,$$

$$\frac{\partial f_1}{\partial x_1}\Big|_o = \frac{\partial f}{\partial x} = f_x, \ldots .$$

$$\vdots$$

Thus, the expansion for f_1 becomes

$$f_1 = f + \alpha U f + \frac{\alpha^2}{2!} U^2 f + \cdots = \sum_{n=0}^{\infty} \frac{\alpha^n}{n!} U^n f = e^{\alpha U} f, \qquad (2.16)$$

where $U^0 \equiv 1$, and the exponential representation is defined by the preceding summation.

Let us set $f_1 = x_1$ and then $f_1 = y_1$ to obtain

$$x_1 = \sum_{n=0}^{\infty} \frac{\alpha^n}{n!} U^n x = \phi(x, y, \alpha),$$

$$y_1 = \sum_{n=0}^{\infty} \frac{\alpha^n}{n!} U^n y = \psi(x, y, \alpha). \qquad (2.17)$$

The above are called the *Lie series* representation of the finite equations of the group. Recall that

$$U^0 x = x, \quad U^0 y = y$$
$$U x = \xi, \quad U y = \eta$$

so that

$$U^2 x = U\xi, \quad U^2 y = U\eta$$

$$\vdots \qquad \qquad \vdots$$

and the Lie series becomes

$$x_1 = x + \alpha\xi + \frac{\alpha^2}{2!} U\xi + \cdots,$$

$$y_1 = y + \alpha\eta + \frac{\alpha^2}{2!} U\eta + \cdots. \qquad (2.18)$$

The finite radius of convergence of the foregoing expansions is of no concern in the subsequent analysis.

As an example, Equation (2.14) is used for the rotation group, to obtain

$$U = -y\frac{\partial}{\partial x} + x\frac{\partial}{\partial y},$$

$$\xi = -y, \quad \eta = x.$$

In view of Equations (2.15), this yields

$$U x = -y, \quad U y = x$$
$$U^2 x = -x, \quad U^2 y = -y$$
$$U^3 x = y, \quad U^3 y = -x$$
$$U^4 x = x, \quad U^4 y = y$$
$$U^5 x = -y, \quad U^5 y = x$$

$$\vdots \qquad \qquad \vdots$$

The sequence repeats every time the index n increases by 4. As a consequence, x_1 is given by Equation (2.17) as

$$x_1 = x - \alpha y - \frac{\alpha^2}{2!}x + \frac{\alpha^3}{3!}y + \frac{\alpha^4}{4!}x - \frac{\alpha^5}{5!}y - \cdots$$

$$= x\left(1 - \frac{\alpha^2}{2!} + \frac{\alpha^4}{4!} - \cdots\right) - y\left(\alpha - \frac{\alpha^3}{3!} + \cdots\right)$$

$$= x\cos\alpha - y\sin\alpha$$

with a similar result for y_1. Equations (2.2) are thereby recovered for the rotation group.

There is a second way of determining the global group equations. Start with the coupled ODEs, which are based on Equations (2.7),

$$\frac{dx_1}{d\alpha} = \xi(x_1, y_1), \quad \frac{dy_1}{d\alpha} = \eta(x_1, y_1), \tag{2.19}$$

and the initial condition

$$x_1 = x, \quad y_1 = y \quad \text{when } \alpha = 0. \tag{2.20}$$

One solution to this system has already been given; namely, the Lie series. An equivalent form, of considerable use in the later development, would be

$$\frac{dx_1}{\xi(x_1, y_1)} = \frac{dy_1}{\eta(x_1, y_1)} = d\alpha. \tag{2.21}$$

Integration of the left equation, which is independent of α, yields a solution that can be written as

$$u(x_1, y_1) = c, \tag{2.22a}$$

where the integration constant c is evaluated when $\alpha = 0$, i.e.,

$$u(x, y) = c \tag{2.22b}$$

in accord with the initial condition. Since these equations are independent of α, they hold for all α. Equation (2.22a) is the solution of the ODE

$$\frac{dy_1}{dx_1} = \frac{\eta(x_1, y_1)}{\xi(x_1, y_1)}, \tag{2.23}$$

and provides the *path curves* of the group. Each distinct c value corresponds to a distinct path curve. Thus, a path curve is a specific solution of this ODE. Suppose that Equation (2.22a) can be solved for x_1 as

$$x_1 = w(y_1, c). \tag{2.24}$$

[Equivalent results are obtained if Equation (2.22a) is solved for y_1.] This is substituted into the rightmost of Equations (2.21)

$$\frac{dy_1}{\eta(x_1, y_1)} = \frac{dy_1}{\eta[w(y_1, c), y_1]} = d\alpha.$$

With x_1 eliminated, integration yields the quadrature solution

$$\int_y^{y_1} \frac{dy_1}{\eta} = \alpha,$$

where the initial condition establishes the integration limits. This solution takes the form

$$\nu(y_1, c) - \nu(y, c) = \alpha.$$

With c given by Equations (2.22), we have

$$\tilde{\nu}(x_1, y_1) - \tilde{\nu}(x, y) = \alpha,$$

where

$$\tilde{\nu}(x, y) = \nu(y, u(x, y)).$$

The global group equations thus have the form

$$u(x_1, y_1) = u(x, y)$$
$$\tilde{\nu}(x_1, y_1) = \tilde{\nu}(x, y) + \alpha. \qquad (2.25)$$

This is a general result that holds for any infinitesimal transformation that satisfies group properties.

Recall that the group property is invariant under any one-to-one change of coordinates. Thus, u and $\tilde{\nu}$ can be viewed as new coordinates. In these new coordinates, the transformation is a simple translation parallel to the $\tilde{\nu}$-axis. The important result is evident that any continuous one-parameter group can be transformed into a translation parallel to one axis.

Once again, the rotation group is used as an example. With Equations (2.10), Equations (2.21) become

$$-\frac{dx_1}{y_1} = \frac{dy_1}{x_1} = d\alpha.$$

The leftmost equation is solved as

$$x_1 dx_1 + y_1 dy_1 = 0$$
$$u(x_1, y_1) = x_1^2 + y_1^2 = c.$$

The rightmost equation becomes

$$\int_y^{y_1} \frac{dy_1}{x_1} = \int_y^{y_1} \frac{dy_1}{(c - y_1^2)^{1/2}} = \int_0^\alpha d\alpha$$
$$\sin^{-1}\left(\frac{y_1}{c^{1/2}}\right) - \sin^{-1}\left(\frac{y}{c^{1/2}}\right) = \alpha,$$

where c can be represented with $x_1^2 + y_1^2$ or $x^2 + y^2$. We drop the tilde notation and set

$$\tan \nu = (y/x),$$

which yields

$$\nu(x, y) = \sin^{-1}\left[\frac{y}{(x^2 + y^2)^{1/2}}\right]$$

and therefore

$$\nu(x_1, y_1) = \nu(x, y) + \alpha.$$

The global equations are recovered by starting from

$$x_1^2 + y_1^2 = x^2 + y^2$$

$$\sin^{-1}\left[\frac{y_1}{(x_1^2 + y_1^2)^{1/2}}\right] - \sin^{-1}\left[\frac{y}{(x^2 + y^2)^{1/2}}\right] = \alpha.$$

The sine is taken of both sides of the second equation, and with the use of

$$\sin\left\{\sin^{-1}\left[\frac{y}{(x^2 + y^2)^{1/2}}\right]\right\} = \frac{y}{(x^2 + y^2)^{1/2}}$$

$$\cos\left\{\sin^{-1}\left[\frac{y}{(x^2 + y^2)^{1/2}}\right]\right\} = \frac{x}{(x^2 + y^2)^{1/2}},$$

Equations (2.2) are finally obtained. Consequently, with the coordinate transformation

$$u = x^2 + y^2, \quad \nu = \tan^{-1}(y/x), \tag{2.26}$$

the rotation group becomes a translation group that is parallel to the ν-axis.

2.4 Problems

2.1. Consider the finite group whose six elements are

$$x, \quad \frac{1}{1-x}, \quad 1 - \frac{1}{x}, \quad \frac{1}{x}, \quad 1 - x, \quad \frac{x}{x-1}$$

where the operation is $f_i \circ f_j = f_i(f_j)$. As an example, we have

$$f_2 \circ f_4 = f_2\left(\frac{1}{x}\right) = \frac{1}{1 - (1/x)} = \frac{x}{x-1} = f_6.$$

Determine the identity element, and find for each f_i its inverse.

2.2. Demonstrate that the following are one-parameter transformation groups:

(1) $x_1 = (x^2 + \alpha xy)^{1/2}, \quad y_1 = \dfrac{xy}{(x^2 + \alpha, xy)^{1/2}},$

(2) $x_1 = x + \alpha$, $y_1 = \dfrac{xy - \alpha}{x + \alpha}$,

(3) $x_1 = \dfrac{x}{1 + \alpha x}$, $y_1 = \dfrac{y}{1 + \alpha x}$.

2.3. Find the infinitesimal transformation $(\delta z, \delta y)$ and the composition law for the transformations in Problem 2.2. Write out the symbol for each transformation.

2.4. Find the infinitesimal transformation and symbol of the following one-parameter groups:

(1) $x_1 = \alpha^2 x$, $y_1 = \alpha^3 y$,

(2) $x_1 = \alpha x + (\alpha - 1)y$, $y_1 = y$.

2.5. For each transformation in Problem 2.2, determine the first three terms of the Lie series representation.

2.6. For each transformation in Problem 2.2, find

$$u = u(x, y),   \nu = \nu(x, y)$$

in accord with Equations (2.25).

2.7. Invert each of the transformations of Problem 2.6, thereby finding $x = x(u, \nu)$ and $y = y(u, \nu)$. What are the group equations of Problem 2.2 in terms of these new coordinates?

2.8. Repeat Problems 2.5, 2.6, and 2.7 for the transformations in Problem 2.4.

2.9. Find the global equations whose infinitesimal transformation are the following:

(1) $x f_x - y f_y$,

(2) $(x + y) f_x$,

(3) $(x - y) f_x + (x + y) f_y$.

Chapter 3

Method of Characteristics

Further progress will require the use of the method of characteristics (MOC). This chapter develops the method and provides several examples. In the next chapter, we return to the theory of continuous groups.

There are many ways to introduce the MOC. For example, Emanuel (1986) provides three different approaches to the subject. Here, interest is limited to a single, first-order, linear or quasilinear PDE. Our MOC approach is thus specifically tailored for the task at hand.

3.1   Theory

For purposes of generality, an inhomogeneous equation is considered for the dependent variable f

$$\sum_{i=0}^{n-1} a_i \frac{\partial f}{\partial x_i} + a_n = 0, \tag{3.1}$$

where n is a positive integer. This equation is assumed to be *quasilinear*, in which case $a_0, \dots, a_n$ can depend on the x_j and f, but not on any derivative of f. We further simplify the equation by noting that if f is a solution, then

$$G(x_o, \dots, x_n) = f(x_o, \dots, x_{n-1}) + x_n \tag{3.2}$$

is a solution of the homogeneous equation

$$\sum_{i=0}^{n} a_i \frac{\partial G}{\partial x_i} = 0. \tag{3.3}$$

Thus, by adding a new independent variable, x_n, the inhomogeneous term is incorporated into the above relation.

Observe that G, when equal to a constant, is a solution of Equation (3.3). This constant may be taken as zero. We, therefore, seek a solution with the

19

form

$$G(x_0, \ldots, x_n) = 0. \tag{3.4}$$

The rest of this section provides this solution.

It is conceptually convenient to introduce an $n + 1$ dimensional Cartesian space that has an orthonormal basis $\hat{|}_j$. Thus, the gradient of G is

$$\nabla G = \sum_{i=0}^{n} \frac{\partial G}{\partial x_i} \hat{|}_i \tag{3.5}$$

and a vector $\vec{A}$ can be defined that is based on the a_i coefficients

$$\vec{A} = \sum_{i=0}^{n} a_i \hat{|}_i.$$

Hence, Equation (3.3) becomes

$$\vec{A} \cdot \nabla G = 0. \tag{3.6}$$

Since i ranges from 0 to n, Equation (3.4) represents a surface in a $n + 1$ dimensional space, and the gradient ∇G is everywhere normal to this surface. On the other hand, $\vec{A}$ is perpendicular to ∇G and therefore $\vec{A}$ is tangent to the surface. Thus, the solution of Equation (3.3) or (3.6) is a surface that is tangent to $\vec{A}$.

Consider a *characteristic curve* that lies on the surface $G = 0$ and which also is everywhere tangent to $\vec{A}$. The surface can be viewed as consisting of an infinite number of these curves. Moreover, each of these curves represents a solution of Equation (3.3).

We next construct a characteristic curve in the $n + 1$ dimensional space whose coordinates are $x_0, \ldots, x_n$. For example, in three dimensions a curve is determined by the intersection of two surfaces. More generally, the characteristic curve we seek is determined by the intersection of the n surfaces

$$u^{(0)}(x_0, \ldots, x_n) = c_0$$
$$u^{(1)}(x_0, \ldots, x_n) = c_1$$

$$\vdots \tag{3.7}$$

$$u^{(n-1)}(x_0, \ldots, x_n) = c_{n-1},$$

where the $u^{(0)}, \ldots, u^{(n-1)}$ are assumed to be functionally independent, the c_j are constants, and the first equation is usually written as $u = c$. Each choice of the c_j yields a different characteristic curve.

Since $\vec{A}$ is tangent to a characteristic curve, the differential change in the x_i along such a curve must stand in the same relation to each other as the corresponding components of $\vec{A}$. Thus, on a characteristic curve, we have

$$\frac{dx_0}{a_0} = \frac{dx_1}{a_1} = \cdots = \frac{dx_n}{a_n}. \tag{3.8}$$

As noted, G is constant along a characteristic curve. We observe from Equation (3.2) that dx_n can be replaced with $-df$. This change is convenient, since the a_i are functions of $x_0, \ldots, x_{n-1}$, and f. The above equations are n coupled, first-order ODEs that relate the x_i along a characteristic curve. The unique solution of these equations is provided by Equations (3.7), where the c_j are the constants of integration. The problem of solving a first-order partial differential equation is thereby reduced to solving n coupled ODEs. As will become apparent, this reduction is advantageous, whether Equation (3.3) is to be solved analytically or numerically.

We now see why Equation (3.3) is limited to being quasilinear. If one of the a_i depended on a derivative of f, then one of Equations (3.8) would not be an ODE, and the theory collapses. Normally, the MOC applies only to hyperbolic equations. For Equation (3.3), this qualification is unnecessary. The only essential restriction is that it be quasilinear.

Note that $\vec{A} \cdot \nabla G$ is also the derivative of G along a characteristic curve. Equation (3.6) therefore means that G has a constant value along any particular characteristic curve. For this to be so, G can depend on the x_i only in combinations such that $dG = 0$ along any characteristic curve. However, the $u^{(j)}$ depend on the x_i, but have a constant value along any characteristic curve. Consequently, G is an arbitrary function of the $u^{(j)}$. The general solution of Equation (3.3) is thus

$$G\left(u^{(0)}, u^{(1)}, \ldots, u^{(n-1)}\right) = 0. \tag{3.9}$$

If one or more of the a_i depend on f, or if $a_n \neq 0$, then f explicitly appears in the $u^{(j)}$, and this equation is also a solution of Equation (3.1). If none of the a_i involve f and $a_n = 0$, then the solution of Equation (3.1) can be written as

$$f = f\left(u^{(0)}, u^{(1)}, \ldots, u^{(n-2)}\right).$$

We verify that Equation (3.9) is a solution of Equation (3.3) by first evaluating $du^{(j)}$ with the aid of Equations (3.8)

$$du^{(j)} = \sum_{i=0}^{n} \frac{\partial u^{(j)}}{\partial x_i} dx_i = \frac{dx_0}{a_0} \sum_{i=0}^{n} a_i \frac{\partial u^{(j)}}{\partial x_i},$$

where we assume that one a_i, say a_0, is nonzero. We next obtain

$$dG = \sum_{j=0}^{n-1} \frac{\partial G}{\partial u^{(j)}} du^{(j)} = \frac{dx_0}{a_0} \sum_{j=0}^{n-1} \frac{\partial G}{\partial u^{(j)}} \sum_{i=0}^{n} a_i \frac{\partial u^{(j)}}{\partial x_i}$$

$$= \frac{dx_0}{a_0} \sum_{i=0}^{n} a_i \sum_{j=0}^{n-1} \frac{\partial G}{\partial u^{(j)}} \frac{\partial u^{(j)}}{\partial x_i} = \frac{dx_0}{a_0} \sum_{i=1}^{n} a_i \frac{\partial G}{\partial x_i} = 0$$

as expected.

The functional form of G is determined by an initial, or boundary, condition. Without loss of generality, this condition may be specified at $x_0 = 0$ as

$$G_0 = G\left[u^{(0)}(0, x_1, \ldots, x_n), \ldots, u^{(n-1)}(0, x_1, \ldots, x_n)\right], \qquad (3.10)$$

where G_0 is the prescribed relation for G at $x_0 = 0$. However, in the subsequent chapters, we will not be concerned with a boundary or initial condition. This is because only the functional form of Equation (3.9) is of interest.

As has been mentioned, the unique solution of Equations (3.8) can be formally written as Equations (3.7). It will be apparent in the next section that an analytical solution of Equations (3.8) often requires inverting some of Equations (3.7). For example, suppose $n = 3$ and we have obtained a solution, $u = c$, to

$$\frac{dx_0}{a_0} = \frac{dx_1}{a_1}.$$

Further, let a_2 depend on x_0, x_1, and x_2. If $u = c$ can be explicitly solved for x_0, we would then integrate

$$\frac{dx_1}{a_1} = \frac{dx_2}{a_2}$$

with x_0 eliminated. Similarly, if $u = c$ is more readily solved for x_1, we might obtain $u^{(1)}$ by integrating

$$\frac{dx_0}{a_0} = \frac{dx_2}{a_2}$$

instead. In either case, the elimination of x_0 (or x_1) from the dx_2 equation is consistent with obtaining a simultaneous solution of Equations (3.8).

3.2   Examples

Let us find the general solution to

$$xz\frac{\partial z}{\partial x} + yz\frac{\partial z}{\partial y} = xy.$$

We first solve the characteristic equations

$$\frac{dx}{xz} = \frac{dy}{yz} = \frac{dz}{xy}.$$

From the leftmost equation, we have

$$\frac{dx}{x} = \frac{dy}{y}$$

which integrates to

$$u = \frac{y}{x} = c.$$

For a second equation, utilize

$$\frac{dx}{z} = \frac{dz}{y}$$

and by elimination of y

$$cx\,dx = z\,dz$$
$$cx^2 = z^2 - c_1$$
$$\left(\frac{y}{x}\right)x^2 = z^2 - c_1$$
$$u^{(1)} = z^2 - xy = c.$$

Hence, the general solution to the PDE is

$$G(z^2 - xy, y/x) = 0,$$

which can be verified by direct substitution. An alternate form for the solution can be written as

$$z^2 - xy = g(y/x),$$

or as

$$z = \pm[xy + g(y/x)]^{1/2},$$

where g is an arbitrary function.

In later chapters, we shall need to find the general solution to

$$Uf = \xi f_x + \eta f_y = 0. \tag{3.11}$$

For this relation, Equations (3.8) become

$$\frac{dx}{\xi} = \frac{dy}{\eta} = \frac{df}{0}$$

and consequently

$$f = \text{constant} \tag{3.12}$$

along the curves provided by

$$\frac{dy}{dx} = \frac{\eta(x, y)}{\xi(x, y)}. \tag{3.13}$$

The solution is written as [see the first of Equations (3.7)]

$$u = u(x, y) = c, \tag{3.14}$$

where c is a constant of integration. This important relation provides the path curves [see Equation (2.23)] of the group, which are the integral curves of $dy/dx = (\eta/\xi)$. If $\xi = 0$, then $u = x = c$, while $u = y = c$ if $\eta = 0$. These two special cases are important for the analysis in Chapter 5.

In view of Equation (3.12), u itself is a solution to $Uf = 0$, as shown by

$$du = \frac{\partial u}{\partial x}dx + \frac{\partial u}{\partial y}dy = u_x dx + u_y \frac{\eta}{\xi}dx$$

$$= (\xi u_x + \eta u_y)\frac{dx}{\xi} = 0.$$

But if u is a solution, so is an arbitrary function of u

$$f(u) = \text{constant}$$

because

$$df = \frac{df}{du}du = \frac{df}{du}\frac{dx}{\xi}(\xi u_x + \eta u_y) = 0.$$

Observe that Equation (3.13) can be written as

$$\frac{dx}{\xi} = \frac{dy}{\eta} = d\alpha, \tag{3.15}$$

which is Equations (2.21). Equation (3.13) is thus equivalent to the symbol of the group and can be referred to as *generating* the group.

In later chapters, we shall also be concerned with a generalization of Equation (3.11), written as

$$U^{(2)}f = \xi f_x + \eta f_y + \eta' f_{y'} + \eta'' f_{y''} = 0, \tag{3.16}$$

where the derivatives

$$y' = \frac{dy}{dx}, \quad y'' = \frac{d^2 y}{dx^2},$$

along with x and y, are considered as the independent variables of f. We also presume the four coefficients are known functions of the form:

$$\begin{aligned}
\xi &= \xi(x, y) \\
\eta &= \eta(x, y) \\
\eta' &= \eta'(x, y, y') \\
\eta'' &= \eta''(x, y, y', y'').
\end{aligned} \tag{3.17}$$

The characteristic equations are

$$\frac{dx}{\xi} = \frac{dy}{\eta} = \frac{dy'}{\eta'} = \frac{dy''}{\eta''}. \tag{3.18}$$

In these equations, y' and y'' are *not* to be replaced by dy/dx and d^2y/dx^2, respectively. The solution of Equation (3.13) is given by Equation (3.14). The solution of

$$\frac{dx}{\xi(x, y)} = \frac{dy'}{\eta'(x, y, y')} \tag{3.19a}$$

requires elimination of y by writing Equation (3.14) as

$$y = w(x, c). \tag{3.20a}$$

This step is unnecessary if Equation (3.19a) is free of y, as sometimes is the case. Similarly, we may use

$$\frac{dy}{\eta} = \frac{dy'}{\eta'} \tag{3.19b}$$

in conjunction with

$$x = w(y, c). \tag{3.20b}$$

[The ws in Equations (3.20) are functionally different.] In either case, the solution is written as

$$u^{(1)}(x, y, y') = c_1, \tag{3.21}$$

where c is eliminated by means of Equation (3.14).

There are three equivalent possibilities for the dy'' equation:

$$\frac{dx}{\xi} = \frac{dy''}{\eta''}, \quad \frac{dy}{\eta} = \frac{dy''}{\eta''}, \quad \frac{dy'}{\eta'} = \frac{dy''}{\eta''}. \tag{3.22}$$

Whichever possibility is chosen generally requires the elimination of x or y and Equation (3.21) to eliminate y' (along with x or y). The solution is written as

$$u^{(2)}(x, y, y', y'') = c_2, \tag{3.23}$$

where c and c_1 are eliminated by means of Equations (3.14) and (3.21).

The general solution of Equation (3.16) is

$$G\left(u, u^{(1)}, u^{(2)}\right) = 0, \tag{3.24}$$

where the $u^{(j)}$ are given by Equations (3.14), (3.21), and (3.23), respectively.

Subsequent chapters often discuss specific examples of Equations (3.11) and (3.16). As will become apparent, the MOC is an essential element in the overall development of the theory.

3.3 Problems

3.1 Find the general solution to

$$xy^2 f_x - x^2 y f_y + \frac{1}{2}(y^2 - x^2)z f_z = 0.$$

3.2 Find the general solution to

$$x f_x - 2f_y - y' f_{y'} - 2y'' f_{y''} = 0.$$

3.3 Find the general solution to

$$(x \ln x) f_x - 2(1+\ln x) f_y - \left[\frac{2}{x} + (1 + \ln x) y'\right] f_{y'} + \left[\frac{2}{x^2} - \frac{y'}{x} - 2(1 + \ln x) y''\right] f_{y''} = 0.$$

3.4 Find the general solution to

$$-y f_x + x f_y + \left[1 + (y')^2\right] f_{y'} + 3 y' y'' f_{y''} = 0.$$

3.5 Find the general solution to

$$(x + y) f_x + (y - x) f_y - [1 + (y')^2] f_{y'} - (1 + 3y') y'' f_{y''} = 0.$$

3.6 Find the general solution to

$$-y f_x + x f_y = 1 + f^2.$$

3.7 Find the general solution to

$$-y f_x + x f_y = (1 - f^2)^{1/2}.$$

3.8 With n constant, find the solution to

$$V_t + V^n V_x = 0$$

that satisfies the initial condition

$$V(x, 0) = V_0(x).$$

Chapter 4

Continuous One-Parameter Groups-II

Although discussed in Section 2.3, the global group equations are of minimal interest. Interest in them is largely limited to showing the connection to the postulates of group theory. By way of contrast, the notion of invariance, discussed in the next section, is of crucial importance.

Up to this point, the concept of a continuous group has been limited to a function $f(x, y)$ of two variables, where x and y transform by means of Equations (2.3). In Section 4.2, we extend the concept to the function $f(x, y, y')$, where y' is the derivative dy/dx and x and y are still governed by Transformation (2.3). A further extension to functions with higher-order derivatives is provided in Section 4.3.

4.1 Invariance

We start with a function $f(x, y)$. The condition under which it is invariant with respect to the group is determined next. By this we mean that

$$f_1 = f(x_1, y_1) = f(x, y) \qquad (4.1)$$

for all the values of the group parameter α. With the aid of Equation (2.16)

$$f(x_1, y_1) = f(x, y) + \alpha U f + \frac{\alpha^2}{2!} U^2 f + \cdots$$

we observe that a necessary and sufficient condition for f to be an invariant function of the group, or for brevity we say f is an *invariant function*, is for

$$U f = 0. \qquad (4.2)$$

All that need be shown is

$$U^n f = 0, n \geq 2$$

27

when $Uf = 0$, but this is obvious. Observe that a global invariance condition is given in terms of the infinitesimal group.

An alternative proof is based on the observation that f_1 satisfies

$$\frac{df_1}{d\alpha} = 0$$

since Equation (4.1) holds for all α. But from Section 2.3, we have

$$\left.\frac{df_1}{d\alpha}\right|_o = Uf,$$

from which Equation (4.2) follows.

As noted in Section 3.2, the characteristic equation for $Uf = 0$ is Equation (3.13). For the rotation group (see Appendix B), this equation integrates to

$$u = x^2 + y^2 = c. \tag{4.3}$$

Hence, any function of u

$$G = G(x^2 + y^2) \tag{4.4}$$

is an invariant function of the rotation group. This means the transformation, which moves x, y in an arc of a circle, does not alter G. The argument of G thus constitutes a path curve [see Equation (2.23)] of Equation (3.13). Alternatively, a path curve transforms into itself under rotation.

The degenerate case when both ξ and η are zero is excluded from consideration. This yields either a fixed point, such as the origin for the rotation group, or a curve consisting of invariant points. For instance, if $\xi = y$ and $\eta = 0$, then $y = 0$ is a curve of invariant points. (Recall that if both ξ and η are identically zero, we do not have a one-parameter group.)

We now state a theorem dealing with the foregoing type of invariance. The curve

$$f(x, y) = 0 \tag{4.5}$$

is invariant under the group if and only if

$$Uf = \omega(x, y)f, \tag{4.6}$$

where ω is an arbitrary function. A more useful alternative statement is

$$Uf = 0 \quad \text{whenever } f = 0. \tag{4.7}$$

This particular invariance formulation is the basis of the analysis in the subsequent chapters. It is understood hereafter that f has a nontrivial dependence on its arguments, i.e., $df \neq 0$. The above condition stems from Equation (4.6). We demonstrate this by first computing

$$U^2 f = (U\omega)f + \omega Uf = (U\omega)f + \omega^2 f = (U\omega + \omega^2)f.$$

Consequently, by induction, all $U^n f$, $n = 1, 2, \ldots$ are proportional to f. Thus, whenever $f = 0$, each term on the right side of Equation (2.16) is zero and f_1 is then zero. However, this is just the condition that f be an invariant function.

Curve Invariance

Another type of invariance is now discussed. Our goal will be to find the condition wherein one curve of a one-parameter family of curves transforms into itself or into another curve of the family.

Let

$$f = f(x, y) = c \tag{4.8}$$

be a one-parameter family of curves. We select one of these curves, which is written as

$$f_1 = f(x_1, y_1) = c_1.$$

Under the Transformation (2.3), this becomes

$$f_1 = f(\phi(x, y, \alpha),\ \psi(x, y, \alpha)) = w(x, y, \alpha) \tag{4.9}$$

or

$$w(x, y, \alpha) = c_1. \tag{4.10}$$

By letting α vary, with c_1 fixed, this relation becomes the other members of the family of curves. Since Equations (4.8) and (4.10) represent the same family of curves for all α, the derivative y' is the same for both equations. It is obtained from

$$df = f_x dx + f_y dy = 0$$
$$y' = -\frac{f_x}{f_y}$$

and from

$$dw = w_x dx + w_y dy = 0$$
$$y' = -\frac{w_x}{w_y}.$$

We thus have

$$w_x f_y - w_y f_x = 0,$$

or

$$\begin{vmatrix} f_x & f_y \\ w_x & w_y \end{vmatrix} = 0.$$

For this determinant to be identically zero, w is a function only of f, or vice versa, i.e.,

$$w = F(f).$$

Consequently, from Equations (4.9) we have

$$f_1 = f(x_1, y_1) = F(f)$$

for all α.

Observe that f_1 is a function of $f(x, y)$, for all values of α, provided each coefficient of α^n on the right side of Equation (2.16) is a function only of f. In particular, we must have

$$Uf = \Omega(f) \qquad (4.11)$$

since higher derivatives, such as

$$U^2 f = U\Omega = \frac{d\Omega}{df} Uf = \Omega \frac{d\Omega}{df}, \qquad (4.12)$$

are functions only of f. With Equation (4.11) satisfied, f_1 is a function of f for all α. This equation provides the invariance condition that any one curve of the family transforms into itself or another member of the family.

We thus have two invariance conditions. Of the two, the one for function invariance is the more important condition. Curve invariance, provided by Equation (4.11), degenerates to function invariance only if $\Omega(f)$ is zero. The following examples, which show how $\Omega(f)$ is evaluated, demonstrate that this is generally not the case.

Examples

An appreciation of the geometrical significance of curve invariance will be useful for the discussion in Section 5.5. As an example, we examine this aspect by considering the family of straight lines (i.e., the elements of the group) that are tangent to a unit circle, as sketched in Figure 4.1. The straight lines are provided by

$$ax + by = 1, \quad a^2 + b^2 = 1$$

or in terms of a single parameter

$$ax + (1 - a^2)^{1/2} y = 1. \qquad (4.13)$$

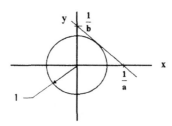

Figure 4.1. Schematic of a straight line that is tangent to a circle of unit radius.

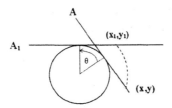

Figure 4.2. Transformation of line A to A_1 as a result of a rotation through an angle θ.

Here, a, or the angle θ, is the group parameter, see Figure 4.2, where changing θ transforms tangent line A into A_1. As θ changes, the line *slides* along the circle. The point of tangency is thus a fixed point on the line. (This statement is established in Section 5.5.) A nontangent point on A, with coordinates x, y, sweeps out a curve shown as the dashed line in Figure 4.2. Since the tangency point on the line is fixed, the dashed curve is an arc of a circle centered about the origin.

The family of straight lines is invariant under a one-parameter rotation group, where $\theta = 0$ is the identity, $\bar{\theta} = -\theta$ is the inverse, and closure is satisfied. The path curves of the group are arcs of circles, such as the dashed curve in Figure 4.2.

The function $\Omega(f)$ is next determined for this particular rotation group. We start by solving Equation (4.13) for a, with the result

$$a = \frac{x + y(x^2 + y^2 - 1)^{1/2}}{x^2 + y^2},$$

where, for convenience, a positive sign is used with the square root. By varying a, the family of straight lines tangent to the unit circle is obtained. Consequently, a is the same as c in Equation (4.8), and f is given by

$$f = \frac{x + y(x^2 + y^2 - 1)^{1/2}}{x^2 + y^2}. \tag{4.14}$$

Its derivatives are

$$f_x = \frac{(-x^2 + y^2)(x^2 + y^2 - 1)^{1/2} - xy(x^2 + y^2 - 2)}{(x^2 + y^2)^2(x^2 + y^2 - 1)^{1/2}}$$

$$f_y = \frac{-2xy(x^2 + y^2 - 1)^{1/2} + x^2(x^2 + y^2 - 1) + y^2}{(x^2 + y^2)^2(x^2 + y^2 - 1)^{1/2}}.$$

Hence, the symbol of the group becomes

$$Uf = -yf_x + xf_y = \frac{-y + x(x^2 + y^2 - 1)^{1/2}}{x^2 + y^2} \tag{4.15}$$

by direct substitution of the derivatives into the rotation symbol. With f given by Equation (4.14), somehow the right side of the above equation must equal $\Omega(f)$.

To find $\Omega(f)$, the derivatives are first evaluated

$$\frac{\partial(Uf)}{\partial x} = \frac{\partial}{\partial x}\left[\frac{-y + x(x^2 + y^2 - 1)^{1/2}}{x^2 + y^2}\right]$$

$$= \frac{2xy(x^2 + y^2 - 1)^{1/2} + y^2(x^2 + y^2 - 1) + x^2}{(x^2 + y^2)^2(x^2 + y^2 - 1)^{1/2}}$$

$$\frac{\partial(Uf)}{\partial y} = \frac{(-x^2 + y^2)(x^2 + y^2 - 1)^{1/2} - xy(x^2 + y^2 - 2)}{(x^2 + y^2)^2(x^2 + y^2 - 1)^{1/2}}.$$

After some algebra, we obtain

$$U^2 f = -y\frac{\partial(Uf)}{\partial x} + x\frac{\partial(Uf)}{\partial y} = -\frac{x + y(x^2 + y^2 - 1)^{1/2}}{x^2 + y^2} = -f.$$

Equation (4.12) is written as

$$U^2 f = \Omega\frac{d\Omega}{df} = -f,$$

or

$$\Omega d\Omega + f df = 0.$$

This integrates to

$$\Omega = \pm\left(k^2 - f^2\right)^{1/2},$$

where k is an integration constant. By substituting this relation and Equation (4.14) into Equation (4.11), we obtain

$$\Omega(f) = \left(1 - f^2\right)^{1/2}$$

for the desired result. Thus, the invariance condition associated with Figures 4.1 and 4.2 is

$$Uf = \left(1 - f^2\right)^{1/2}.$$

This particular example is reexamined in Section 5.5.

A simpler example is provided by the family of straight lines

$$y = ax$$

that pass through the origin. Under rotation, a line transforms into a second line, say,

$$y = bx.$$

In this case, one can show that Equation (4.11) becomes

$$Uf = -yf_x + xf_y = 1 + f^2.$$

4.2   The Once-Extended Group

Let Transformation (2.3) define a one-parameter group in two variables. We consider the derivative

$$y' = \frac{dy}{dx} \tag{4.16}$$

as a third variable to be transformed by the group. However, y' is not an independent variable; its group equation is given by

$$y_1' = \frac{dy_1}{dx_1} = \frac{d\psi}{d\phi} = \frac{\frac{\partial \psi}{\partial x}dx + \frac{\partial \psi}{\partial y}dy}{\frac{\partial \phi}{\partial x}dx + \frac{\partial \phi}{\partial y}dy} = \frac{\psi_x + \psi_y y'}{\phi_x + \phi_y y'} = \theta(x, y, y', \alpha). \tag{4.17}$$

One can show that adding $y_1' = \theta$ to the group equations for x and y results in a system of equations that maintain the group property. For instance, the identity transformation has

$$\phi = x, \quad \psi = y,$$

which results in

$$\phi_x = 1, \quad \phi_y = 0, \quad \psi_x = 0, \quad \psi_y = 1,$$

and

$$y_1' = \frac{0 + y'}{1 + 0y'} = y'$$

as was to be shown.

The new group is called the *once-extended group*. As was done for x_1 and y_1, we need to determine $\delta y_1'$, subject to Equation (4.16), and the symbol for the extended group.

We utilize the Lie series to obtain

$$dx_1 = dx + \alpha d\xi + \cdots$$
$$dy_1 = dy + \alpha d\eta + \cdots.$$

For an infinitesimal change, $\delta\alpha, y_1'$ becomes

$$y_1' = \frac{dy_1}{dx_1} = \frac{dy + \delta\alpha d\eta}{dx + \delta\alpha d\xi} = \frac{y' + \frac{d\eta}{dx}\delta\alpha}{1 + \frac{d\xi}{dx}\delta\alpha}$$

$$= \left(y' + \frac{d\eta}{dx}\delta\alpha\right)\left(1 - \frac{d\xi}{dx}\delta\alpha + \cdots\right)$$

$$= y' + \left(\frac{d\eta}{dx} - y'\frac{d\xi}{dx}\right)\delta\alpha + O(\delta\alpha)^2.$$

For an infinitesimal transformation, we thus have

$$\delta y' = y_1' - y' = \eta'\delta\alpha,$$

where

$$\eta' \equiv \frac{\delta y'}{\delta \alpha} = \frac{d\eta}{dx} - y'\frac{d\xi}{dx} \tag{4.18a}$$

$$= \left(\frac{\partial \eta}{\partial x} + \frac{\partial \eta}{\partial y}y'\right) - y'\left(\frac{\partial \xi}{\partial x} + \frac{\partial \xi}{\partial y}y'\right)$$

$$= \eta_x + (\eta_y - \xi_x)y' - \xi_y(y')^2. \tag{4.18b}$$

These equations provide the variation in y' with α such that Equation (4.16) holds. They are the y' counterpart to Equations (2.7). Observe that the primes on y and η have different meanings. It is best to regard the prime on η as purely notational.

Next, consider an infinitesimal change in a function f of x, y, and y' as follows:

$$f_1 = f(x_1, y_1, y_1') = f(x, y, y') + \frac{\partial f}{\partial x}\delta x + \frac{\partial f}{\partial y}\delta y + \frac{\partial f}{\partial y'}\delta y' + \cdots$$

$$\delta f = f_1 - f = f_x \xi \delta \alpha + f_y \eta \delta \alpha + f_{y'} \eta' \delta \alpha + O(\delta \alpha)^2.$$

We define the symbol, $U^{(1)}$, of the once-extended group as

$$U^{(1)}f \equiv \frac{\delta f}{\delta \alpha} = \xi f_x + \eta f_y + \eta' f_{y'} = Uf + \eta' f_{y'},$$

where η' is given by Equation (4.18b). Thus, $f(x, y, y')$ is an invariant function of the once-extended group if and only if

$$U^{(1)}f = 0 \quad \text{whenever } f = 0. \tag{4.19}$$

This represents a straightforward generalization of Equations (4.7).

For the convenience of the reader, key formulas are listed in Appendix C, where the operational definitions of η' and $U^{(1)}f$ can be found.

For the rotation group, Appendix B, Equation (4.18b) yields

$$\eta' = 1 + (0 - 0)y' - (-1)(y')^2 = 1 + (y')^2$$

and the once-extended symbol is

$$U^{(1)}f = -yf_x + xf_y + [1 + (y')^2]f_{y'}.$$

4.3   Higher-Order Extended Groups

The results in the preceding section are generalized. We start with Transformation (2.3) and form the derivatives, with α held constant,

$$y_1' = \frac{dy_1}{dx_1} = \frac{d\psi}{d\phi} = \frac{\psi_x + \psi_y y'}{\phi_x + \phi_y y'} = \theta(x, y, y', \alpha) \tag{4.17}$$

$$y_1'' = \frac{dy_1'}{dx_1} = \frac{d\theta}{d\phi} = \frac{\theta_x + \theta_y y' + \theta_{y'} y''}{\phi_x + \phi_y y'} = \omega(x, y, y', y'', \alpha) \tag{4.20}$$

$$\vdots$$

The symbol for the nth-extended group is defined as

$$U^{(n)}f = \xi f_x + \eta f_y + \eta' f_{y'} + \eta'' f_{y''} + \cdots + \eta^{(n)} f_{y^{(n)}} \qquad (4.21)$$

where

$$\xi = \frac{\delta x}{\delta \alpha}, \quad \eta = \frac{\delta y}{\delta \alpha}$$

$$\eta' = \frac{\delta y'}{\delta \alpha} = \frac{d\eta}{dx} - y' \frac{d\xi}{dx}$$

$$\eta'' = \frac{\delta y''}{\delta \alpha} = \frac{d\eta'}{dx} - y'' \frac{d\xi}{dx} \qquad (4.22)$$

$$\vdots$$

$$\eta^{(n)} = \frac{\delta y^{(n)}}{\delta \alpha} = \frac{d\eta^{(n-1)}}{dx} - y^{(n)} \frac{d\xi}{dx}.$$

(Observe that $U^{(n)}f$ is different from the $U^n f$ of Section 4.1.) The last relation represents a recursive formula for $\eta^{(n)}$ based on the operator

$$\frac{d}{dx} = \frac{\partial}{\partial x} + y' \frac{\partial}{\partial y}.$$

This relation provides an efficient means for generating the $\eta^{(n)}$ coefficients. For example, with η' given by Equation (4.18b), we have for η'':

$$\begin{aligned}
\eta'' &= \frac{d}{dx}[\eta_x + (\eta_y - \xi_x)y' - \xi_y(y')^2] - y''(\xi_x + \xi_y y') \\
&= \eta_{xx} + \eta_{xy}y' + (\eta_{xy} + \eta_{yy}y' - \xi_{xx} - \xi_{xy}y')y' + (\eta_y - \xi_x)y'' \\
&\quad - (\xi_{xy} + \xi_{yy}y')(y')^2 - 2\xi_y y'y'' - (\xi_x + \xi_y y')y'' \qquad (4.23) \\
&= \eta_{xx} + (2\eta_{xy} - \xi_{xx})y' + (\eta_{yy} - 2\xi_{xy})(y')^2 - \xi_{yy}(y')^3 \\
&\quad + (\eta_y - 2\xi_x - 3\xi_y y')y''.
\end{aligned}$$

Although η' is quadratic in y', see Equation (4.18b), $\eta^{(n)}$ is linear in $y^{(n)}$ for all $n \geq 2$. This is evident for η'' in the above equation.

4.4 Problems

4.1. Determine η' and η'' for the transformations in Problem 2.2.

4.2. Determine η' and η'' for the transformations in Problem 2.4.

4.3. Determine η' and η'' for the symbols in Problem 2.9.

4.4. The one-parameter family

$$(x - a)^2 - y^2 = a^2$$

is invariant under the group

$$Uf = xf_x + \frac{x^2}{y}f_y.$$

Determine the function $\Omega(f)$ in Equation (4.11).

4.5. Find the symbol, η', and η'' for the group

$$x_1 = xe^{m\alpha}, \quad y_1 = ye^{n\alpha}$$

where m and n are constants.

4.6. The uniform dilatation group

$$x_1 = xe^{\alpha}, \quad y_1 = ye^{\alpha}$$

has curve invariance for circles whose center is at the origin. Determine ξ, η, and $\Omega(f)$ that appears in Equation (4.11).

Part II

ORDINARY DIFFERENTIAL EQUATIONS

Chapter 5

First-Order ODEs

We are now in a position to apply the theory in Part I to an ODE. This chapter considers the simplest case of a first-order equation. However, a given ODE may not be invariant under any transformation group, in which case this approach is of little use. In fact, one can show that the number of groups a first-order ODE is invariant under may range from none to an infinite number. (This topic is further discussed in Section 6.1.) When the number is two or more, the choice of which group to use for a solution is then a matter of convenience.

Suppose we have an ODE, of any order, that is invariant under a group as represented by a specific symbol Uf. Section 2.1 showed that group properties are invariant under any one-to-one transformation. In other words, any such transformation of the dependent and independent variables results in a transformed ODE that is still invariant under the group. Because there is an infinitude of possible transformations, the form of the ODE that is invariant under the group is of some generality.

The central idea is to use an inverse approach in which we start with a known symbol. First, the general form for the first-order ODE

$$f(x, y, y') = 0 \qquad (5.1)$$

is found that is invariant under the group represented by the symbol. (Function, rather than curve, invariance is invoked.) Our objective is to find the general solution to this ODE. This equation is generally nonlinear and nonseparable and involves at least one arbitrary function. Second, again starting with the known symbol, a transformation of coordinates is determined that replaces Equation (5.1) with a separable form. This relation is then integrated, after which the original variables are reintroduced. The procedure for developing the transformation is discussed in Section 5.2; this procedure is free of any guesswork. Thus, an exact analytical solution of Equation (5.1) is established in terms of quadratures. In brief, any first-order ODE that is invariant under a known group can be analytically solved.

A variety of special procedures or generalizations are discussed in Section 5.3. These significantly extend the class of first-order ODEs amenable to a group-theory solution.

Because there is no systematic procedure for finding the group under which a given first-order ODE is invariant, further progress requires the above inverse approach. We thus assume a group and find the general form for the first-order ODE under which it is invariant. In this manner, a compendium of ODEs invariant under various groups is set forth in Section 5.4. Given a first-order ODE, we first see if it fits any of the tabulated ones. If it does, a separable solution in terms of quadratures is assured.

Section 5.5 provides examples that illustrate the theory. Higher-order ODEs invariant under a one-parameter group are the subject of Chapter 6.

A very useful feature of the group approach is that the ensuing theory does not depend on the singularities of the differential equation. Since a knowledge of the type and location of the singularities is not needed, this aspect is not discussed. Similarly, there is no explicit discussion of initial conditions.

5.1 Invariance Under a One-Parameter Group

We assume the symbol of the group is known, in which case the symbol of the once-extended group $U^{(1)}f$ is readily found. Starting with Uf is preferable to starting with the global equations. In fact, if we started with Transformation (2.3), the first step would be to find the symbol, after which the global equations are no longer of interest.

The most general form for a first-order ODE can be written as Equation (5.1). We assume this ODE is invariant under a group whose symbol is Uf. Since f involves y', f is an invariant function of the once-extended group. In other words,

$$U^{(1)}f = \xi f_x + \eta f_y + \eta' f_{y'} = 0 \tag{5.2}$$

whenever Equation (5.1) holds.

A first-order ODE is often written as

$$y' = g(x, y). \tag{5.3}$$

This form is less general than Equation (5.1), but suffices for the vast majority of cases, and is often used. Nevertheless, the theory is based on Equation (5.1). Our first objective is to solve Equation (5.2), thereby determining a specific functional form for Equation (5.1), which is then an invariant function with respect to the group whose symbol is Uf.

Equation (5.2) is solved by the MOC, so that

$$\frac{dx}{\xi} = \frac{dy}{\eta} \tag{5.4}$$

and

$$\frac{dy'}{\eta'} = \frac{dx}{\xi}, \tag{5.5a}$$

or

$$\frac{dy'}{\eta'} = \frac{dy}{\eta}. \tag{5.5b}$$

As before, let $u(x, y) = c$ be the solution to Equation (5.4). In accord with Equation (4.2), the solution u will be referred to as the *invariant* of the group. The function, $f = u(x, y) - c$, is then an invariant of both $Uf = 0$ and $U^{(1)}f = 0$. In other words, $Uu(x, y) = 0$ and $U^{(1)}u(x, y) = 0$, since u is independent of y'.

A second function

$$u^{(1)} = u^{(1)}(x, y, y') = c_1 \tag{5.6}$$

is required that is a solution of either Equation (5.5a) or (5.5b). This function is called the *first differential invariant* and, in contrast to u, explicitly depends on y'.

A general solution of Equation (5.2) is sought. In accord with Equation (3.9), it is given by

$$G\left(u, u^{(1)}\right) = 0, \tag{5.7a}$$

or by

$$u^{(1)} = g(u), \tag{5.7b}$$

where G and g are arbitrary functions. To obtain u and $u^{(1)}$, two ODEs need to be solved. For the theory to be useful, Equations (5.5) must differ from Equation (5.1) and one of them should be solvable. Both conditions are generally fulfilled.

Usually, Equation (5.4) is easily solved, often by inspection. Equations (5.5) are frequently much more complicated in appearance. Nevertheless, these equations possess general, closed-form solutions, which are derived in the next subsection. The following section, however, will show that $u^{(1)}$, in fact, is quite unnecessary for a solution.

Riccati Equation

With the aid of Equation (4.18b), Equation (5.5a) is written as

$$\frac{dy'}{dx} = \frac{\eta'}{\xi} = \frac{\eta_x}{\xi} + \left(\frac{\eta_y - \xi_x}{\xi}\right) y' - \frac{\xi_y}{\xi} (y')^2. \tag{5.8}$$

This is a *Riccati equation* for y' with known coefficients on the right side that depend only on x. Any y dependence is presumed to be removed with $u(x, y) = c$. The Riccati equation is nonlinear, and there is no general solution for it. However, a general solution (Rainville, 1943) does exist when a particular solution is known. We next show that w', which stems from Equation (3.20a), is a particular solution. This step is not trivial, since $y = w(x, c)$ is obtained as a solution of Equation (5.4), not Equation (5.5a).

As mentioned, the various coefficients on the right side of the above equation depend only on x, since y is eliminated by means of Equation (3.20a). By differentiation, we have

$$y' = w' = \frac{\eta}{\xi}.$$

A second differentiation yields

$$\frac{d^2w}{dx^2} = \frac{dw'}{dx} = \frac{1}{\xi^2}\left(\xi\frac{d\eta}{dx} - \eta\frac{d\xi}{dx}\right) = \frac{1}{\xi}(\eta_x + \eta_y y') - \frac{\eta}{\xi^2}(\xi_x + \xi_y y').$$

Replace both y' and η/ξ with w' to obtain

$$\frac{dw'}{dx} = \frac{\eta_x}{\xi} + \left(\frac{\eta_y - \xi_x}{\xi}\right)w' - \frac{\xi_y}{\xi}(w')^2. \tag{5.9}$$

This equation, however, is the same as Equation (5.8). Thus, w' is a known particular solution of this equation.

The general solution of the Riccati equation is temporarily denoted as $\tilde{y}'$ in order to avoid confusing it with the particular solution w', which equals y'. To obtain the general solution, set

$$\tilde{y}' = w' + \frac{1}{z} \tag{5.10}$$

which introduces the new dependent variable z. By differentiation, we obtain

$$\frac{d\tilde{y}'}{dx} = \frac{dw'}{dx} - \frac{1}{z^2}\frac{dz}{dx}.$$

Replace $d\tilde{y}'/dx$ with Equation (5.8) and dw'/dx with Equation (5.9), with the result

$$-\frac{1}{z^2}\frac{dz}{dx} = \left(\frac{\eta_y - \xi_x}{\xi}\right)(\tilde{y}' - w') - \frac{\xi_y}{\xi}\left[(\tilde{y}')^2 - (w')^2\right].$$

Next, replace $\tilde{y}'$ with Equation (5.10) and w' with η/ξ to obtain

$$-\frac{1}{z^2}\frac{dz}{dx} = \left(\frac{\eta_y - \xi_x}{\xi}\right)\frac{1}{z} - \frac{\xi_y}{\xi}\left(\frac{2\eta}{z\xi} + \frac{1}{z^2}\right).$$

After simplification, this equation becomes

$$\frac{dz}{dx} + Pz = Q, \tag{5.11}$$

where

$$P(x) = \frac{\eta_y - \xi_x}{\xi} - \frac{2\xi_y\eta}{\xi^2}, \quad Q(x) = \frac{\xi_y}{\xi}. \tag{5.12}$$

Equation (5.11) is a first-order, linear ODE in standard form whose general solution is

$$z = \exp\left(-\int P\,dx\right)\left\{c_1 + \int Q\left[\exp\left(\int P\,dx\right)\right]dx\right\}, \tag{5.13}$$

where c_1 is an integration constant. [This solution if often utilized in the later analysis.] We thus have for y', where the tilde notation is dropped,

$$y' = w' + \frac{1}{z} = \frac{\eta}{\xi} + \frac{\exp(\int P dx)}{c_1 + \int Q \exp(\int P dx) dx}.$$

The η/ξ term depends only on x and y, while the several integrands depend only on c and x, since Equation (3.20a) is used to eliminate y. After the integrations are performed, c is replaced by $u(x, y)$. Solve for the constant c_1 to obtain the first differential invariant as

$$u_{Rx}^{(1)} = u_{Rx}^{(1)}(x, y, y') = c_1, \tag{5.14a}$$

where x, y, and y' are independent variables, and

$$u_{Rx}^{(1)} = \frac{\xi \exp(\int P dx)}{\xi y' - \eta} - \int Q \exp\left(\int P dx\right) dx. \tag{5.14b}$$

Note the appearance of $\xi y' - y$ in the denominator. This factor is not zero; it is the only place that y' occurs, and this factor will often appear in the subsequent theory.

There are several forms for $u^{(1)}$. To distinguish them, an R subscript, for Riccati, is used. The x subscript means Equation (5.5a) is utilized in the Riccati equation; a y subscript would appear when Equation (5.5b) is used. No subscript is used on $u^{(1)}$ when the Riccati equation procedure is not directly utilized. Much of the time, it is unnecessary to distinguish between the various forms for $u^{(1)}$. However, this notation will prove useful for clarifying the subsequent discussion.

When η' is a complicated function and $u(x, y) = c$ is most easily solved for y, i.e., $y = w(x, c)$, then $u^{(1)}$ is best obtained from Equation (5.14b). The results of this procedure are summarized in Appendix C. This appendix also provides the comparable relations for $u_{Ry}^{(1)}$ when $u(x, y) = c$ is most readily solved for x.

When $\eta \equiv 0$, we have

$$u = y = c$$

and Equation (5.14b) simplifies to

$$u_{Rx}^{(1)} = \frac{1}{y'} e^{-\int \frac{\xi_x}{\xi} dx} - \int \frac{\xi_y}{\xi} e^{-\int \frac{\xi_x}{\xi} dx} dx.$$

In the integrands, y is held constant, with the result

$$\int \frac{\xi_x}{\xi} dx = \int \frac{d\xi}{\xi} = \ln \xi(x, y)$$

and $u_{Rx}^{(1)}$ becomes

$$u_{Rx}^{(1)} = \frac{1}{\xi y'} - \int \frac{\xi_y}{\xi^2} dx.$$

Similarly, when $\xi \equiv 0$, we have

$$u = x = c$$

and

$$u_{Ry}^{(1)} = \frac{y'}{\eta} - \int \frac{\eta_x}{\eta^2} dy.$$

Equivalence

There are a number of *equivalent forms* for $u^{(1)}$. The equivalence concept for the differential invariants is important in the subsequent analysis and needs to be made precise. Two otherwise different differential invariants, say $\tilde{u}^{(1)}$ and $u^{(1)}$, are equivalent if a relation of the form

$$F\left(u, u^{(1)}, \tilde{u}^{(1)}\right) = 0 \tag{5.15}$$

exists. Consequently, Equation (5.7a) becomes

$$G\left(u, \tilde{u}^{(1)}\right) = 0.$$

Alternatively, $u^{(1)}$ and $\tilde{u}^{(1)}$ are equivalent if both

$$U^{(1)}u^{(1)} = 0, \quad U^{(1)}\tilde{u}^{(1)} = 0$$

are satisfied. Section 6.1 has a more extensive discussion of this topic and provides a proof of this last assertion. The choice of form for the first differential invariant is usually a matter of convenience, e.g., the simplest among equivalent forms might be preferred.

Rotation Group

As an illustration, we again resort to the rotation group. The invariant u is given by Equation (4.3). With Appendix B, Equation (5.5b) becomes

$$\frac{dy}{x} = \frac{dy}{(c - y^2)^{1/2}} = \frac{dy'}{1 + (y')^2},$$

which integrates to

$$\sin^{-1}\left(\frac{y}{c^{1/2}}\right) = \tan^{-1} y' + \tan^{-1} c_1.$$

Since

$$\frac{y}{c^{1/2}} = \frac{y}{(x^2 + y^2)^{1/2}},$$

we have

$$\sin^{-1}\left(\frac{y}{(x^2 + y^2)^{1/2}}\right) = \tan^{-1}\left(\frac{y}{x}\right). \tag{5.16}$$

After simplification, the first differential invariant is

$$u^{(1)}(x, y, y') = \frac{y - xy'}{x + yy'} = c_1 \tag{5.17}$$

for the rotation group.

To illustrate the Riccati equation approach for $u_{Rx}^{(1)}$, Appendix B is utilized for the rotation group, with the result:

$$P(x) = \frac{2x}{y^2} = \frac{2x}{c - x^2}, \quad Q(x) = \frac{1}{y} = \frac{1}{(c - x^2)^{1/2}}$$

$$\int P dx = 2 \int \frac{x dx}{c - x^2} = -\ln(c - x^2)$$

$$\exp\left(\int P dx\right) = \frac{1}{c - x^2} = \frac{1}{y^2}$$

$$\int Q \exp\left(\int P dx\right) dx = \int \frac{dx}{(c - x^2)^{3/2}} = \frac{x}{c(c - x^2)^{1/2}} = \frac{x}{y(x^2 + y^2)}$$

$$\frac{\xi \exp(\int P dx)}{\xi y' - \eta} = -\frac{y(y)^{-2}}{-yy' - x} = \frac{1}{y(yy' + x)}.$$

Equation (5.14b) thus becomes

$$u_{Rx}^{(1)} = \frac{1}{y(yy' + x)} - \frac{x}{y(x^2 + y^2)} = \frac{1}{x^2 + y^2}\frac{y - xy'}{x + yy'} = \frac{u^{(1)}}{u}.$$

Hence, $u^{(1)}$ and $u_{Rx}^{(1)}$ are equivalent to each other.

The general form for a first-order ODE invariant under rotation then is

$$G(u, u^{(1)}) = G\left(x^2 + y^2, \frac{y - xy'}{x + yy'}\right) = 0. \tag{5.18a}$$

Alternatively, we can write

$$\frac{y - xy'}{x + yy'} = g(x^2 + y^2) \tag{5.18b}$$

or, upon solving for y',

$$y' = \frac{y - xg(x^2 + y^2)}{x + yg(x^2 + y^2)}. \tag{5.18c}$$

The last form is often the most convenient one, although all three forms are considered equivalent to each other.

Discussion

The path curves are a one-parameter family, given by $u(x, y) = c$, where c is the parameter. This family of curves should not be confused with the *integral curves* that are solutions of the ODE, Equation (5.7a) or (5.7b). (Actually, the $u = c$ path curves are also integral curves, but for the $y' = \eta/\xi$ ODE.) The integral curves are also a one-parameter family, whose parameter is the constant of integration. The connection between the two families is indicated in Figure 5.1, where they generate a nonorthogonal grid. The integral curves are solutions of a first-order ODE that is invariant under the group whose path curves, $u = c$, are shown. Consider a particular integral curve whose constant of integration is a_1. If the value of the constant of integration is changed to a_2, a new integral curve is obtained, as shown in the figure. In going continuously from a_1 to a_2, point 1 on the a_1 integral curve will move along a specific path curve to point 2, which is on the a_2 integral curve. We see that invariance means that any point on an integral curve moves along a specific *path curve* as the integration constant changes. This aspect is further discussed in Section 5.4 when solutions of Equations (5.18) are considered.

As might be expected, there is a connection between group methods and finding an *integrating factor*, $\mu(x, y)$. Suppose the ODE

$$M(x, y)dx - N(x, y)dy = 0 \tag{5.19}$$

is invariant under Equation (2.12), and the path curves of this equation do not coincide with the integral curves of the above equation, i.e.,

$$\xi M - \eta N \neq 0.$$

Then an integrating factor of Equation (5.19) is

$$\mu = (\xi M - \eta N)^{-1}.$$

Table 5.1 in Appendix D lists the symbol, group description, and ODE invariant under the group for a few groups. This table is deliberately brief and is included only for purposes of discussion. A more comprehensive table is provided in Section 5.4. In accord with Equations (5.7), the ODE is given in two alternative forms. While the once-extended symbol is not shown, it can be found from Uf.

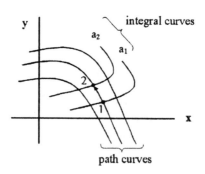

Figure 5.1. Schematic of the path and integral curves.

Several patterns are evident in the table. Observe that u, which is the first argument of G, is a solution of Equation (3.13) that does not depend on y'. As seen by Equation (5.6), the y' derivatives enters only through $u^{(1)}$, which is the second argument of G, where y' appears in the first power. Thus, u and $u^{(1)}$ are independent of each other. If Equation (5.7a) can be written as Equation (5.7b), the ODE can then be explicitly written with a y' on the left side, as is done in Equation (5.18c) and in Table 5.1. This step is not always possible. For example, if G has the form

$$G = \ln u^{(1)} + \exp u^{(1)} + u = 0,$$

then Equation (5.7b) cannot be explicitly written. Observe that only for the first four cases in Table 5.1 can the ODE be solved directly by separation of variables. This topic is further discussed in the next section.

It is possible to obtain an erroneous result when $\xi = 1$. (A similar difficulty occurs when $\eta = 1$.) In this case, Equation (5.5a) becomes

$$dx = \frac{dy'}{\eta'}$$

and η' is given by

$$\eta' = \frac{d\eta}{dx} - y'\frac{d\xi}{dx} = \frac{d\eta}{dx}$$

because $\xi = 1$. Eliminate η' to obtain

$$dy' = d\eta,$$

which integrates to

$$u^{(1)} = y' - \eta(x, y) = c_1.$$

This relation, however, is generally not correct for the first differential invariant. The reason for this is that the $dy' = d\eta$ equation should be written as

$$\frac{dy'}{dx} = \eta_x + \eta_y y',$$

which usually requires the use of $y = w(x, c)$ for its solution. With $\xi = 1$, the correct form for the first differential invariant can be obtained from Equation (5.14b), with the result

$$u_{Rx}^{(1)} = (y' - \eta)\exp\left(-\int \eta_y dx\right) = c_1.$$

The two invariants are in accord only if

$$\int \eta_y dx$$

is a function of u, where u is a solution of $y' = \eta$. This condition is generally not met, as can be verified by direct calculation using an assumed relation for $\eta(x, y)$.

5.2   Canonical Coordinates

The ODEs for items 5 and 6 in Table 5.1 cannot be solved directly by separation
of variables, as can the first four. However, Section 2.3 shows that any one-
parameter group can be transformed by a change of variables into a translation
group, which would correspond to cases 1 or 2 in Table 5.1. This result is
utilized to show that we can always transform, in a variety of ways, an invariant
ODE so that it can be solved by separation of variables. This result is achieved
by introducing *canonical variables* or *coordinates*, denoted as X and Y.

Let us consider the transformation

$$X = X(x, y), \quad Y = Y(x, y) \tag{5.20}$$

and its inverse

$$x = x(X, Y), \quad y = y(X, Y), \tag{5.21}$$

where the Jacobian of the transformation is assumed not to equal zero. We first
determine the symbol

$$U f(X, Y) = \xi(x, y)\frac{\partial}{\partial x}f(X, Y) + \eta(x, y)\frac{\partial}{\partial y}f(X, Y)$$

$$= \xi\left(\frac{\partial f}{\partial X}\frac{\partial X}{\partial x} + \frac{\partial f}{\partial Y}\frac{\partial Y}{\partial x}\right) + \eta\left(\frac{\partial f}{\partial X}\frac{\partial X}{\partial y} + \frac{\partial f}{\partial Y}\frac{\partial Y}{\partial y}\right)$$

$$= (\xi X_x + \eta X_y)f_X + (\xi Y_x + \eta Y_y)f_Y. \tag{5.22}$$

A transformed symbol operator is defined as

$$\overline{U} f(X, Y) = \overline{\xi}(X, Y)f_X + \overline{\eta}(X, Y)f_Y \tag{5.23}$$

where, by comparison with Equation (5.22),

$$\overline{\xi} = \xi X_x + \eta X_y, \quad \overline{\eta} = \xi Y_x + \eta Y_y. \tag{5.24}$$

Although x and y in the arguments of ξ and η can be replaced by Transformation
(5.21), it is convenient not to do so. As in Equations (2.15), we have

$$\overline{U}X = \overline{\xi}, \quad \overline{U}Y = \overline{\eta}, \tag{5.25}$$

in accord with Equations (5.23) and (5.24).

Equation (5.1), which admits a one-parameter group, is transformed into a
separable form

$$\frac{dY}{dX} = g(X) \tag{5.26}$$

that corresponds to a translation parallel to the Y-axis. According to item 1 in
Table 5.1, this will occur if

$$\overline{U}f = f_Y,$$

which directly implies

$$\bar{\xi} = 0, \quad \bar{\eta} = 1.$$

These relations constitute the transformation equations. Written out, $\bar{\xi} = 0$ is

$$\xi(x,y)X_x + \eta(x,y)X_y = 0.$$

The solution of this equation is the invariant of the group, $u(x,y) = c$. We thus set

$$X = u(x,y). \tag{5.27}$$

Since $u = c$, u is only determined to within a nonzero multiplicative constant. Consequently, the right side of the above equation can be replaced with a constant times u, should this prove convenient.

The second condition, $\bar{\eta} = 1$, leads to

$$\xi(x,y)Y_x + \eta(x,y)Y_y = 1$$

where the MOC solution is given by

$$\frac{dx}{\xi} = \frac{dy}{\eta} = dY. \tag{5.28}$$

Since the second coordinate must be different from X, we must use the dY term, thereby leading to two alternate possibilities

$$Y = \int \frac{dy}{\eta[w(y,c),y]}, \tag{5.29a}$$

$$Y = \int \frac{dx}{\xi[x,w(x,c)]}, \tag{5.29b}$$

where the two w functions are different, but both stem from $u(x,y) = c$. The constant c is replaced with u after the integration. The use of Equations (3.20) to eliminate x or y does not contradict the requirement that the Y transformation equation be independent of that for X. Generally, this x or y elimination is required by Equation (3.13), which holds for both Y canonical variables.

Discussion

A number of special topics are discussed. To begin, suppose η is a function only of y; then Equation (5.29a) reduces to

$$Y = \int \frac{dy}{\eta(y)}$$

and the use of $u = c$ to eliminate x is unnecessary. A similar result occurs with Equation (5.29b) if ξ is only a function of x. This type of simplification also occurs if either ξ or η is zero. For instance, if $\xi = 0$, then $u = x = c$ and Equation (5.29a) becomes

$$Y = \int \frac{dy}{\eta(c, y)}$$

with a similar result when $\eta = 0$.

The derivative

$$dY = Y_x dx + Y_y dy$$

is often needed in the subsequent analysis. It can be obtained directly from Equations (5.28) if ξ is only a function of x or η is only a function of y. Otherwise, one of Equations (5.29) is utilized. In doing so, it is important to remember that c is a function of x and y. Thus, the integration should be performed and c replaced with $u(x, y)$ before Y_x and Y_y are evaluated. For example, let

$$Uf = [A(x) + B(y)]f_x$$

then

$$\eta = 0, \quad X = u = y = c$$

and

$$Y = \int \frac{dx}{\xi} = \int \frac{dx}{A(x) + B(c)}.$$

This integral, of course, can be evaluated only after $A(x)$ is prescribed. Now, if $A = x^2$ and $B > 0$, we have

$$Y = \int \frac{dx}{x^2 + B(c)} = \frac{1}{[B(y)]^{1/2}} \tan^{-1} \left\{ \frac{x}{[B(y)]^{1/2}} \right\}$$

and the partial derivatives are

$$Y_x = \frac{1}{x^2 + B(y)}, \quad Y_y = -\frac{B'}{2B^{3/2}} \left[\tan^{-1} \left(\frac{x}{B^{1/2}} \right) + \frac{xB^{1/2}}{x^2 + B} \right].$$

More generally,

$$Y = \int \frac{dy}{\eta(x, y)}, \quad Y = \int \frac{dx}{\xi(x, y)}$$

directly yields

$$dY = -dx \int \frac{\partial \eta}{\partial x} \frac{dy}{\eta^2} + \frac{dy}{\eta} \tag{5.30a}$$

and

$$dY = \frac{dx}{\xi} - dy \int \frac{\partial \xi}{\partial y} \frac{dx}{\xi^2}, \tag{5.30b}$$

where $x\,(y)$ is replaced with $w(y, c)\,[w(x, c)]$ in the integrands and c is replaced with $u(x, y)$ after the integration. With, e.g.,

$$\xi = x^2 + B(y), \quad \eta = 0$$

Equation (5.30b) yields the Y_x and Y_y values of the preceding example.

The integrals in Equations (5.30) should not be evaluated by an integration by parts. For instance, consider the case

$$\xi = A(xy), \quad \eta = 0$$

with

$$X = u = y = c \tag{5.31a}$$

$$Y = \int \frac{dx}{\xi} = \int \frac{dx}{A(cx)}. \tag{5.31b}$$

Equation (5.30b) can be written as

$$dY = \frac{dx}{A} - \frac{dy}{y^2} \int z \frac{dA}{A^2}, \tag{5.32a}$$

where $z = xy$, and in the integrand we have $z = cx$ and $dz = cdx$. If e.g., $A = xy$, then the above relation results in

$$dY = \frac{dx}{xy} - \frac{(\ln x)dy}{y^2}, \tag{5.32b}$$

which can also be directly obtained from Equation (5.31b). On the other hand, an integration by parts for the integral in Equation (5.32a) yields

$$\int z \frac{dA}{A^2} = -\frac{z}{A} + \int \frac{dz}{A} = -\frac{xy}{A} + y \int \frac{dx}{A} = -\frac{xy}{A} + yY.$$

Because of the xy/A term, this result does not correspond to Equation (5.32b) when A is xy. It appears that the uv term in the integration by parts should be dropped.

There are other possible canonical transformations that lead to separable forms as indicated by the second through fourth items in Table 5.1. In each instance, the invariant of the group can be chosen as one coordinate, and there are two possibilities for the second coordinate. Which of the possibilities is chosen might depend on the ease of performing the various quadratures.

The g functions in Equations (5.7b) and (5.26) are arbitrary, but not the same. With this in mind, these equations can be written as

$$u^{(1)} = g_1(u) = g_1(X)$$
$$\frac{dY}{dX} = g(X).$$

Thus, dY/dX is equivalent to $u^{(1)}$, and Equation (5.26) is an alternative form for Equation (5.7b). Similarly,

$$\frac{dY}{dX} = g(Y) \tag{5.33}$$

is another alternative form, with dY/dX equivalent to $u^{(1)}$, when

$$Y = u(x, y)$$

and

$$X = \int \frac{dy}{\eta}, \quad X = \int \frac{dx}{\xi}.$$

It is, therefore, not essential to evaluate $u^{(1)}$. Instead, Equations (5.27) and (5.29), or the above, are directly used for X and Y after which dY/dX is found. This is usually the simplest approach for finding the ODE that is invariant under the group.

Example

Let us transform the rotation group ODE to the canonical form of Equation (5.26). From Equation (5.18a),

$$X = x^2 + y^2$$

and Equations (5.29) result in

$$Y = \int \frac{dy}{\eta} = \int \frac{dy}{x} = \int \frac{dy}{(c - y^2)^{1/2}} = \sin^{-1}\left(\frac{y}{c^{1/2}}\right) = \sin^{-1}\left[\frac{y}{(x^2 + y^2)^{1/2}}\right]$$

$$Y = \int \frac{dx}{\xi} = -\int \frac{dx}{(x - x^2)^{1/2}} = -\sin^{-1}\left[\frac{x}{(x^2 + y^2)^{1/2}}\right].$$

Choosing either of these yields

$$dY = \frac{x\,dy - y\,dx}{x^2 + y^2} = -\frac{y - xy'}{x^2 + y^2}dx, \tag{5.34a}$$

while dX is given by

$$dX = 2x\,dx + 2y\,dy = 2(x + yy')dx.$$

This produces the desired result

$$\frac{dY}{dX} = -\frac{1}{2X}\frac{y - xy'}{x + yy'} = \frac{1}{X}g(X) \tag{5.34b}$$

in view of Equation (5.18b), where the $-(1/2)$ is absorbed in the g function.

The transformation can be simplified by using Equation (5.16) to obtain

$$X = x^2 + y^2, \quad Y = \tan^{-1}\left(\frac{y}{x}\right) \tag{5.35}$$

which also yields Equation (5.34b). On the other hand, the variable change, $Y = (y/x)$, does not lead to a separable form.

5.3 Special Procedures

One difficulty with the group approach is the limited number of ODEs that fall within its scope. This limitation is partly rectified by considering techniques that increase the scope of the theory. The first of these procedures is called the *substitution principle*.

Substitution Principle

This principle is based on the following interchange:

$$\xi \rightleftharpoons \eta, \quad x \rightleftharpoons y, \quad y' \rightleftharpoons 1/y', \tag{5.36}$$

where the derivative substitution stems from

$$\frac{dy}{dx} \rightarrow \frac{dx}{dy} = \frac{1}{y'}.$$

The interchange replaces $\xi f_x + \eta f_y$ with $\eta f_y + \xi f_x$, which leaves Uf unaltered. (In other words, if $Uf = x^m y^n$, then the substitution principle yields $Uf = x^n y^m$.) Consequently, Equation (5.1) becomes

$$f(y, x, 1/y') = 0. \tag{5.37}$$

The principle holds for the canonical coordinates; hence, X becomes $u(y, x)$ and Equations (5.29a) and (5.29b) are interchanged. Thus, the derivative relation

$$\frac{dY}{dX} = \frac{y' - \eta \int \frac{\partial \eta}{\partial x} \frac{dy}{\eta^2}}{u_x(\eta - \xi y')}, \tag{5.38a}$$

which is based on Equation (5.29a), becomes by the substitution principle

$$\frac{dY}{dX} = -\frac{1 - \xi y' \int \frac{\partial \xi}{\partial y} \frac{dx}{\xi^2}}{u_y(\eta - \xi y')}. \tag{5.38b}$$

This relation can also be directly obtained from Equation (5.29b).

As indicated by Equation (5.37), the substitution principle may generate a distinct alternative form for the ODE that is invariant under the group. This is not always the case, as can be seen by applying the principle to Equation (5.18a), wherein the ODE transforms into itself. The principle does transform the Riccati equation for $u_{Rx}^{(1)}$ into the one for $u_{Ry}^{(1)}$, and vice versa. Both results are summarized in Appendix C.

Interchange Principle

The canonical coordinates, X and Y, can also be *interchanged*, inasmuch as Y can equal u and the integrals in Equations (5.29) can equal X. For example, Equation (5.26) becomes

$$\frac{dX}{dY} = g(Y),$$

or

$$\frac{dY}{dX} = g(Y), \tag{5.33}$$

where the two gs are functionally different. This equation corresponds to a translation parallel to the x-axis, and can be obtained by using $\bar{\xi} = 1$ and $\bar{\eta} = 0$ with the transformation procedure of Section 5.2. This X, Y interchange is useful, since it may result in a more convenient form for g. Both the substitution principle and the interchange procedure yield the original ODE when repeated twice.

General Forms for the Symbol

Franklin (1928) provides three generic symbols that can be used to establish other, more specific, symbols. This list is not inclusive, but does encompass many of the first-order ODEs known to be invariant under a group. The symbols and their corresponding ODEs and canonical coordinates are provided in Table 5.2. The various functions $A(x), \ldots, Q(x, y), E(Q)$ are arbitrary, and P and Q should not be confused with the P and Q in Equations (5.12). For the three types in the table, the canonical coordinates yield Equation (5.26).

For type I in the table and Equation (5.26), we have

$$\xi = AB, \quad \eta = 0, \quad \bar{\xi} = 0, \quad \bar{\eta} = 1 \tag{5.39}$$

and u satisfies

$$\xi u_x + \eta u_y = AB u_x = 0.$$

Hence, u is a function only of y, say $u = y$, and Equation (5.27) yields

$$X = u = y.$$

For Y, Equation (5.29b) is utilized, with $y = c$,

$$Y = \int \frac{dx}{\xi} = \int \frac{dx}{A(x)B(c)} = \frac{1}{B(y)} \int \frac{dx}{A}.$$

[Equation (5.29a) cannot be used, since $\eta = 0$.] For $u^{(1)}$, Equation (5.14b) can be used, as follows:

$$P = -\frac{A'}{A}, \quad Q = \frac{B'}{B}$$

$$\int P dx = -\ln A$$

$$u_{Rx}^{(1)} = \frac{1}{Ay'} - \int \frac{B'(c)\,dx}{B(c)}\frac{dx}{A} = \frac{1}{Ay'} - \frac{B'}{B}\int \frac{dx}{A}.$$

Thus, Equation (5.7a) has the form shown in Table 5.2.
As an example of type I, suppose we have

$$A = x^a, \quad B = \exp\left[(a-1)\int P(y)dy\right], \tag{5.40}$$

where a is a constant not equal to unity, and $P(y)$ is an arbitrary function. We thus have

$$\frac{1}{Ay'} - \frac{B'}{B}\int \frac{dx}{A} = \frac{1}{x^a y'} + (1-a)P\frac{x^{1-a}}{1-a} = \frac{1}{x^a y'} + x^{1-a}P$$

and the ODE invariant under the group is

$$\frac{1}{y'} + xP(y) = x^a g(y).$$

The substitution principle is now used to obtain the more familiar form

$$y' + P(x)y = g(x)y^a, \tag{5.41}$$

which is the (Jacob) *Bernoulli equation*. Equation (5.11) is readily obtained by setting $a = 0$.

So far, the focus has been on deriving a general form for the ODE that is invariant under a given group. In doing so, canonical coordinates are obtained. Equations (5.26) and (5.33) are readily integrated

$$Y = a + \int g(X)dx, \quad X = a + \int \frac{dY}{g(Y)}, \tag{5.42}$$

where a is an integration constant. With the Transformation (5.20), the general solution of the ODE is then obtained in terms of the original x, y variables.

The ODEs shown in Table 5.2 are rather general, and some specialization is necessary for practical applications, as indicated by the Bernoulli equation example. The process of generating a general solution of Equation (5.41), in terms of quadratures, is now demonstrated. The most difficult aspect is relating the P and g functions that appear in the equation to canonical coordinates. Keep in mind that $a \neq 1$ and that Equation (5.41) stems from the substitution principle.

From Table 5.2, we have

$$X = x, \quad Y = \frac{1}{B(x)} \int \frac{dy}{A(y)}$$

and

$$dX = dx, \quad dY = -\frac{B'}{B^2} dx \int \frac{dy}{A} + \frac{dy}{AB} = \left(-\frac{B'}{B} Y + \frac{y'}{AB} \right) dx$$

$$\frac{dY}{dX} = \frac{y'}{AB} - \frac{B'}{B} Y = g_1(X).$$

By comparing the rightmost equation with Equation (5.41), we obtain

$$P(x)y = -AB'Y, \quad g(x)y^a = ABg_1,$$

where the $g(x)$ equation readily yields

$$A(y) = y^a, \quad g(x) = B(x)g_1(X).$$

Hence, we obtain

$$Y = \frac{1}{B} \int \frac{dy}{y^a} = \frac{1}{(1-a)B(x)y^{a-1}}$$

and

$$P(x) = -\frac{A}{y} B'Y = -\frac{y^a}{y} B' \frac{1}{(1-a)By^{a-1}} = \frac{B'}{(a-1)B}.$$

The above A and P relations are in accord with Equations (5.40). As a consequence, the canonical transformation equations are

$$X = x, \quad Y = \frac{e^{(1-a)\int P dx}}{(1-a)y^{a-1}}$$

and Equation (5.26) has the form

$$\frac{dY}{dX} = g_1(X) = \frac{g(X)}{B(x)} = g(X)e^{(1-a)\int P dx}.$$

The separation of variable solution then is

$$Y = b + \int g(X)e^{(1-a)\int P dx} dX,$$

where b is an integration constant. In terms of the original variables, we have

$$\frac{1}{(1-a)B(x)y^{a-1}} = b + \int g(x)e^{(1-a)\int P dx} dx$$

or

$$y = \left\{ (1-a)B(x) \left[b + \int g(x)e^{(1-a)\int P dx} dx \right] \right\}^{1/(1-a)},$$

where B is provided by Equation (5.40). This result can be verified as the solution of Bernoulli's equation by differentiation.

For type II in Table 5.2, X stems from

$$\frac{dx}{A} = \frac{dy}{BC}$$

which yields

$$X = u = \int \frac{B}{A} dx - \int \frac{dy}{C}$$

while Y comes from Equation (5.29b)

$$Y = \int \frac{dx}{\xi} = \int \frac{dx}{A}.$$

The Riccati equation requires

$$P = \frac{B}{A} \frac{dC}{dy} - \frac{A'}{A} = \left(\frac{dC}{C} - \frac{dA}{A} \right) \frac{1}{dx}, \quad Q = 0$$

and yields

$$u_{Rx}^{(1)} = \left(\frac{A}{C} y' - B \right)^{-1}.$$

As an example, suppose

$$A = ax, \quad B = -b, \quad C = y$$

where a and b are constants. We then obtain

$$X = \int \frac{B}{A} dx - \int \frac{dy}{C} = -\ln(x^{b/a} y) = -\frac{1}{a} \ln(x^b y^a)$$

and

$$\frac{A}{C} y' - B = \frac{ax}{y} y' + b.$$

The ODE invariant under the group is

$$\frac{ax}{y} y' + b = g \left[-\frac{1}{a} \ln(x^b y^a) \right].$$

A new g function can be defined as

$$g(x^b y^a) = -\frac{b}{a} + \frac{1}{a} g \left[-\frac{1}{a} \ln(x^b y^a) \right]$$

to yield the more elegant result

$$xy' = yg(x^b y^a). \tag{5.43}$$

Although the above two gs are functionally different, it is sometimes notationally convenient not to distinguish between them. This practice, already adhered to, is often used.

For Equation (5.43), with Table 5.2 and

$$A = ax, \quad B = -b, \quad C = y,$$

the transformation is

$$X = -\frac{1}{a}\ln(x^b y^a), \quad Y = \frac{1}{a}\ln x.$$

Equation (5.26) becomes

$$dX = -\left(\frac{y'}{y} + \frac{b}{ax}\right)dx, \quad dY = \frac{dx}{ax}$$

$$\frac{dY}{dX} = -\frac{y}{axy' + by} = g_1(X).$$

We thus have

$$xy' = -\frac{y}{ag_1(X)} - \frac{b}{a}y = yg(x^b y^a) = yg(e^{-aX}).$$

By comparison with Equation (5.03), this yields

$$g_1(X) = -\frac{1}{b + ag(e^{-aX})}.$$

The separation of variable solution then is

$$Y = c + \int g_1(X)dX,$$

where c is a constant of integration. In terms of the original variables, we can write

$$\ln x = ac - a\int \frac{dX}{b + ag(e^{-aX})},$$

where X is replaced by its transformed value after the integration is performed.

In both the type I and II examples, two ODEs are compared after solving for identical terms that contain y'. This quite typical step enables us to identify the various arbitrary functions that can appear in the ODEs.

For type III, X stems from

$$\frac{dx}{P_y} = -\frac{dy}{P_x}$$

$$P_x dx + P_y dy = 0$$

$$X = u = P.$$

For Y, we use

$$Y = \int \frac{dx}{\xi} = \int \left(Q_x - \frac{P_x}{P_y} Q_y \right) \frac{dx}{E} = \int \frac{1}{E} (Q_x dx + Q_y dy) = \int \frac{dQ}{E(Q)}.$$

The Riccati equation is too complicated for obtaining $u^{(1)}$. Instead, we use

$$\frac{dY}{dX} = \frac{\dfrac{dQ}{E(Q)}}{dP} = \frac{1}{E} \frac{Q_x dx + Q_y dy}{P_x dx + P_y dy} = \frac{1}{E} \frac{Q_x + Q_y y'}{P_x + P_y y'}.$$

Since Equation (5.26) holds, the right side equals $g(X)$, which also equals $g(P)$. Thus, the right side is equivalent to $u^{(1)}$, which establishes the $G = 0$ result in Table 5.2.

As an example, suppose

$$P = \ln \left(\frac{y}{x^n} \right), \quad Q = \ln \left(x^a y^b \right), \quad E = e^Q,$$

where a, b, and n are constants. We readily obtain

$$P_x = -\frac{n}{x}, \quad P_y = \frac{1}{y}, \quad Q_x = \frac{a}{x}, \quad Q_y = \frac{b}{y}, \quad E = x^a y^b,$$

which results in

$$\frac{1}{x^a y^b} \frac{(a/x) + (b/y)y'}{-(n/x) + (1/y)y'} = g \left[\ln \left(\frac{y}{x^n} \right) \right].$$

This relation simplifies to

$$x^a y^b \frac{xy' - ny}{bxy' + ay} = g \left(\frac{y}{x^n} \right). \tag{5.44}$$

The substitution principle can be applied to the symbols in Table 5.2. For type II, the procedure yields a symbol of the form

$$Uf = A(y)f_x + B(x)C(y)f_y.$$

In contrast to the original type II symbol, this symbol does not provide a useful result. For instance, X and Y are given by

$$X = u = \int B \, dx - \int \frac{A}{C} \, dy$$

$$Y = \int \frac{dx}{A(y)} \quad \text{or} \quad \int \frac{dy}{B(x)C(y)}.$$

While the X relation is alright, the Y quadratures cannot be performed because $u = c$ cannot generally be solved for x or for y. This difficulty does not arise with the symbols listed in Table 5.2.

Enlargement Procedure

The final topic of this section is called the *enlargement procedure*. Imagine starting with a known symbol Uf and the corresponding u and $u_{Rx}^{(1)}$ invariants. A new symbol of the form

$$\widehat{U}f = h(x,y)Uf \qquad (5.45)$$

is defined, where $\widehat{U}f$ represents the enlarged group. When h is a nonzero constant, we will show that the ODE invariant under the original group is unchanged. When h is not a constant, a new ODE is obtained that is invariant under the enlarged group but not invariant under the original group.

Because h multiplies Uf, Equation (5.4) is unaltered and the two groups have the same invariant function u. Thus, only $u_{Rx}^{(1)}$ and $\widehat{u}_{Rx}^{(1)}$ differ. We proceed to determine how these two differential invariants are related.

For the enlarged group, we have

$$\widehat{\xi} = h\xi, \quad \widehat{\eta} = h\eta$$

so that Equations (5.12) and (5.14b) become

$$\widehat{P}(x) = P(x) - \frac{1}{\xi h}Uh, \quad \widehat{Q}(x) = Q(x) + \frac{h_y}{h}$$

$$\widehat{u}_{Rx}^{(1)} = \frac{\xi \exp(\int \widehat{P}dx)}{\xi y' - \eta} - \int \widehat{Q}\exp\left(\int \widehat{P}dx\right)dx \qquad (5.46)$$

where ξ, η, P, and Q are associated with Uf. When h is a nonzero constant, P and Q equal $\widehat{P}$ and $\widehat{Q}$, respectively. Consequently, $\widehat{u}_{Rx}^{(1)}$ equals $u_{Rx}^{(1)}$ and the invariant ODE is unchanged.

For purposes of simplicity, one might assume that h is a function only of x or of y. A much more useful and interesting result, however, is obtained by assuming h is an arbitrary function of u, i.e.,

$$h = p(u). \qquad (5.47)$$

Henceforth, this relation is assumed throughout the rest of the book. In this circumstance, we have

$$Uh = Up(u) = p'Uu = 0$$

$$\widehat{P} = P, \quad \widehat{Q} = Q + \frac{p'}{p}u_y,$$

where

$$p' = \frac{dp}{du},$$

and

$$\widehat{u}_{Rx}^{(1)} = u_{Rx}^{(1)} - \int \frac{p'}{p}u_y e^{\int Pdx}dx$$

in place of Equation (5.46). In the integral, y is replaced with $w(x, c)$, when necessary. There is, of course, a dy-integrand version, which stems from the use of $u_{Ry}^{(1)}$.

The extended symbol for the enlarged group can be written as

$$\widehat{U}^{(1)}f = pUf + \widehat{\eta}' f_{y'},$$

where $\widehat{\eta}'$ equals

$$\widehat{\eta}' = p\eta' + \frac{u_x}{\eta}p'(\eta - \xi y')^2$$

$$= p\eta' - \frac{u_y}{\xi}p'(\eta - \xi y')^2.$$

The first $\widehat{\eta}'$ form should be used if $\xi = 0$, while the second form is used if $\eta = 0$. If $\xi\eta \neq 0$, use either form. Thus, the extended symbol becomes

$$\widehat{U}^{(1)}f = pU^{(1)}f + \frac{u_x}{\eta}p'(\eta - \xi y')^2 f_{y'},$$

or with $-u_y/\xi$ replacing u_x/η. Since p', by assumption, is not zero, the coefficient of $f_{y'}$ is not identically zero. As a consequence, the ODE that satisfies

$$U^{(1)}f = 0$$

does not satisfy

$$\widehat{U}^{(1)}f = 0,$$

and vice versa.

The analysis is simplified by not evaluating $u^{(1)}$ or $\widehat{u}^{(1)}$. To proceed, we note that the ODE invariant under the enlarged group can be transformed into a separable form

$$\frac{d\widehat{Y}}{d\widehat{X}} = \widehat{g}(\widehat{X}). \tag{5.48}$$

The corresponding canonical coordinates are

$$\widehat{X} = X = u$$
$$\widehat{Y} = \int \frac{dx}{p(u)\xi} = \frac{1}{p(u)}\int \frac{dx}{\xi} = \frac{Y}{p(X)}, \tag{5.49}$$

where Y is given by either of Equations (5.29), and u is a constant inside the integrand. With the derivative

$$\frac{d\widehat{Y}}{d\widehat{X}} = \frac{1}{p}\frac{dY}{dX} - \frac{Y}{p^2}\frac{dp}{dX},$$

Equation (5.48) can be written as

$$\frac{dY}{dX} + Yq(X) = g(X), \tag{5.50}$$

where

$$q = -\frac{1}{p}\frac{dp}{dX}, \quad g = p\widehat{g}. \tag{5.51}$$

Alternatively, p is given by

$$p(u) = \exp\left(-\int q(u)du\right)$$

since $X = u$. Both p and $\widehat{g}$ are arbitrary functions, and, therefore, so are q and g, which have the same X argument. An explicit form for $p(u)$ can be obtained only if an explicit form for $q(X)$ is prescribed. In the subsequent theory, this is not done, and no attempt is made to evaluate $p(u)$.

Note that X and Y are canonical coordinates for the original ODE but not for the enlarged one. Similarly, $\widehat{X}$ and $\widehat{Y}$ are canonical coordinates for the enlarged ODE but not for the original one. It is nevertheless convenient to write the ODE in terms of X and Y, as is done in Equation (5.50). While this equation is not separable, unless q or g is zero, it is linear.

The interchange procedure, i.e.,

$$\widehat{Y} = u, \quad \widehat{X} = \frac{X}{p(Y)}$$

holds, with the result

$$\frac{dX}{dY} + Xq(Y) = g(Y). \tag{5.52}$$

This relation is sometimes more convenient than is Equation (5.50).

Equations (5.50) and (5.52) have the same linear form as Equation (5.11). With the aid of Equations (5.12) and (5.13), the general solutions are

$$Y = \exp\left(-\int qdX\right)\left[c_1 + \int g\exp\left(\int qdX\right)dX\right] \tag{5.53a}$$

$$X = \exp\left(-\int qdY\right)\left[c_1 + \int g\exp\left(\int qdY\right)dY\right], \tag{5.53b}$$

where c_1 is a constant of integration. After the quadratures are performed, X and Y are replaced by means of Equations (5.20). The result is the solution of the enlarged ODE.

Equations (5.50) and (5.52) are invariant under the $\widehat{U}f$ group. They respectively reduce to Equations (5.26) and (5.33) when $p = 1$ or $q = 0$. Equations (5.50) and (5.52), however, are more general because they involve two arbitrary functions rather than one. Earlier ODEs sometimes involved more than one arbitrary function. Type I in Table 5.2, e.g., has $A(x)$ and $B(y)$, which must be prescribed. This is not the case for g and q. One of the virtues of Equations (5.50) and (5.52) is that $p(u)$, and therefore $\widehat{U}f$, need not be known or determined.

As an example of the foregoing discussion, let us start with

$$\xi = y^n, \quad \eta = 0,$$

where n is a real number. (This example is item Ib1 in Table 5.4.) We readily obtain

$$X = u = y, \quad Y = \frac{x}{y^n}$$

and Equation (5.26) becomes

$$\frac{dY}{dX} = \frac{1 - \frac{nx}{y}y'}{y^n y'} = g_1(X) = g_1(y).$$

For purposes of clarity in this example, it is necessary to keep track of the different gs and qs. Upon simplification, the above ODE becomes

$$\frac{1}{y'} - \frac{nx}{y} = y^n g_1 = g_2(y).$$

With

$$U = y^n \frac{\partial}{\partial x}$$
$$\eta' = \eta_x + (\eta_y - \xi_x)y' - \xi_y(y')^2 = -ny^{n-1}(y')^2,$$

the extended symbol is

$$U^{(1)} = y^n \frac{\partial}{\partial x} - ny^{n-1}(y')^2 \frac{\partial}{\partial y'}.$$

We write

$$f = \frac{1}{y'} - \frac{nx}{y} - g_2(y), \quad f_x = -\frac{n}{y}, \quad f_{y'} = -\frac{1}{(y')^2}$$

to obtain

$$U^{(1)}f = 0,$$

as expected.

For the enlargement, Equation (5.50) is utilized, to obtain

$$\frac{1 - \frac{nx}{y}y'}{y^n y'} + \frac{x}{y^n}q(y) = g(y),$$

or

$$\frac{1}{y'} - \frac{nx}{y} + xq = y^n g.$$

After simplification, this becomes

$$\frac{1}{y'} + x\tilde{q}(y) = \tilde{g}(y),$$

where

$$\tilde{q} = q - \frac{n}{y} = -\frac{p'}{p} - \frac{n}{y}, \quad \tilde{g} = y^n g = y^n p g_1.$$

Since $u = y$, an explicit form for $p(u)$ can be obtained only if an explicit form for $\tilde{q}(y)$ is provided. This is not done, and thus $p(u)$ remains unknown. Nevertheless, we demonstrate that $\hat{U}^{(1)} f$ equals zero. We start by evaluating

$$\hat{\eta}' = p\eta' - \frac{u_y}{\xi} p'(\eta - \xi y')^2$$

$$= -ny^{n-1}(y')^2 p - \frac{p'}{y^n}(y^n y')^2$$

$$= -y^n (y')^2 \left(\frac{n}{y} p + p'\right)$$

and the extended symbol is

$$\hat{U}^{(1)} f = py^n f_x - y^n (y')^2 \left(\frac{n}{y} p + p'\right) f_{y'}.$$

With

$$f = \frac{1}{y'} + x\tilde{q}(y) - \tilde{g}(y)$$

$$f_x = \tilde{q} = -\frac{p'}{p} - \frac{n}{y}, \quad f_{y'} = -\frac{1}{y'^2},$$

we have

$$\hat{U}^{(1)} f = -py^n \left(\frac{p'}{p} + \frac{n}{y}\right) + y^n \left(\frac{n}{y} p + p'\right) = 0,$$

as was to be shown.

The foregoing enlarged ODE becomes

$$y' + y\tilde{q}(x) = \tilde{g}(x)$$

with the substitution principle. To check this, we start with the transformed infinitesimal elements

$$\xi = 0, \quad \eta = x^n.$$

These yield

$$X = u = x, \quad Y = \int \frac{dy}{\eta} = \frac{y}{x^n}, \quad \frac{dY}{dX} = \frac{1}{x^n}\left(y' - \frac{ny}{x}\right)$$

and Equation (5.50) becomes

$$\frac{1}{x^n}\left(y' - \frac{ny}{x}\right) + \frac{y}{x^n} q(x) = g(x).$$

Upon simplification, we have

$$y' + y\left[q(x) - \frac{n}{x}\right] = x^n g(x),$$

which is in accord with the above ODE. This verifies that the substitution principle holds for ODEs based on the enlargement procedure.

Observe that a given symbol may correspond to as many as four distinct ODEs, each of which contains two arbitrary (enlargement) functions. Two of the ODEs stem from the substitution principle, while two stem from the interchange procedure.

5.4 Compendium

Starting with a particular one-parameter group, the ODE invariant under it can be found. We examine the converse in the following subsection.

PDE Formulation

Suppose the ODE we wish to solve is $y' = g(x, y)$. Then, Equation (5.2) is

$$\xi g_x + \eta g_y - \eta' = 0,$$

where f is given by

$$f = g(x, y) - y'.$$

Equations (4.18b) and (5.3) are used to obtain

$$\xi g_x + \eta g_y = \eta_x + (\eta_y - \xi_x)g - \xi_y g^2. \tag{5.54}$$

If the group is given, this is a quasilinear PDE for g. On the other hand, suppose g is known, then there is only a single equation for ξ and η, and the system is *underdetermined*. We could set ξ, or η, equal to zero, which results in a linear PDE for η, or ξ, which is solvable with the MOC. In this case, the ODE is invariant under an infinite number of Lie groups. Of course, only a single group is needed for the solution of $y' = g$. For instance, if we set ξ or η equal to zero, the MOC solution then yields $y' = g$, which is the original ODE whose solution we had hoped to obtain.

As an example, suppose

$$y' = g = \phi(x)\psi(y)$$

and let us correctly guess that $\xi = 0$. Equation (5.54) becomes

$$\eta_x + \phi\psi\eta_y = \phi\psi'\eta,$$

which has the solution

$$\eta = \psi(y).$$

However, if we start with $\xi = xy$, we then have to solve

$$\eta_x + \phi\psi\eta_y = \phi\psi'\eta + xy\psi\phi' + y\phi\psi + x\phi^2\psi^2.$$

Using the MOC, we obtain

$$\frac{dx}{1} = \frac{dy}{\phi\psi} = \frac{d\eta}{\phi\psi'\eta + xy\psi\phi' + y\phi\psi + x\phi^2\psi^2},$$

and the leftmost equation is simply $y' = g(x, y)$, while the equation involving $d\eta$ cannot be integrated in closed form.

Nevertheless, Ovsiannikov (1983) has proposed solving Equation (5.54) with the substitution

$$\eta = \xi g + \theta(x, y),$$

where θ is a new unknown function of x and y. This substitution results in

$$\theta_x + g\theta_y = g_y\theta$$

for Equation (5.54). Aside from the trivial solution, $\theta = 0$, the characteristic equations are

$$\frac{dx}{1} = \frac{dy}{g} = \frac{d\theta}{g_y}.$$

Again, the leftmost equation is just $y' = g(x, y)$ for which we seek a solution.

The foregoing approach is therefore useless for establishing a group when g is given. This is not the case when the ODE is of second or higher order, as will be discussed in the next chapter. In this circumstance, the ODE has only a finite number of groups under which it is invariant, and the system of PDEs for ξ and η are overdetermined.

ODE Catalogue

In view of the foregoing discussion, a practical approach would be to tabulate ODEs for known groups, as is done in a table of integrals. Table 5.3 outlines such a compendium for first-order ODEs. In the table, η is divided into five categories, while ξ is divided into four. Systematic use is made of the substitution principle and the interchange procedure. Consequently, a symbol may generate as many as four distinct ODEs. Table 5.3 is used in conjunction with Table 5.4, which provides specific results for a given symbol. The item column in Table 5.4 provides the symbol type, an s when the substitution principle is utilized, and a consecutive number designation. The enlargement column, for a given item, provides, in order, results based on Equations (5.50) and (5.52). The only exceptions are when a result is redundant or excessively complicated. Because of the substitution principle, items IIIb, IVa, and IVb are unnecessary, while type V covers those cases that don't fit into the first four categories. A compendium could have been developed around the generic symbols in Table 5.2. The current approach, however, is more comprehensive and should be easier to implement with computer software.

In Table 5.4, there are a number of arbitrary functions, such as $A, B, M, \ldots$, and constants, such as $a, b, m, n, \ldots$. Constraints are occasionally present in the symbol column, e.g., see items IC2 and IC3. These are necessary, e.g., when separately dealing with polynomial and logarithm results when integrating a quantity such as x^m. Both the $m = -1$ and $m \neq -1$ cases are covered. The arguments of the arbitrary functions are provided in the symbol column.

The third column lists the invariant function of the group, which equals X in accord with Equation (5.27). The next two columns provide Y and dY/dX. While X and Y may be simplified with respect to multiplicative constants, the derivative is based on the tabulated X and Y values without any further simplification. By including dY/dX, there is no need to evaluate $u^{(1)}$. The resulting table is more compact and convenient, since listing $u^{(1)}$ does not remove the necessity for determining both Y and dY/dX.

The dY/dX derivative was discussed in Section 5.2; see also Equations (5.38). In terms of the table, however, it is based on

$$\frac{dY}{dX} = \frac{Y_x + Y_y y'}{X_x + X_y y'}$$

whenever explicit algebraic relations for $X(x, y)$ and $Y(x, y)$ are available. When $X = c$, the denominator can be written as

$$X_x + X_y y' = \begin{cases} \dfrac{X_y}{\xi}(\xi y' - \eta) \\ -\dfrac{X_x}{\eta}(\xi y' - \eta). \end{cases}$$

Thus, only X_x or X_y need be evaluated. The Y_x and Y_y partial derivatives can stem from Equation (5.29), after integration, or from Equations (5.32).

The usefulness of the table is significantly enhanced by including one or two ODEs for each enlarged group. While duplicate ODEs are usually avoided, straightforward generalizations are retained; compare items Ib1 and Ic2. Both ODEs are simplified as much as possible. In fact, the first enlargement for Ib1 can be simplified to the point where $N(y)$ can be absorbed into the arbitrary $q(y)$ and $g(y)$ functions. This is not the case for the second Ib1 enlargement. The same thing occurs for the two Ibs1 enlargements.

In the enlargement column, q and g have the same argument, which is only shown for q. When q is zero, the ODE is comparable to Equations (5.26) or (5.33), and separation of variables is applicable. This yields a solution of the form of Equations (5.42).

Most often, q is not zero, and the linear ODE has the solution provided by Equations (5.53). Although dY/dX does not appear in these equations, the derivative is nevertheless essential for evaluating q and g, since these functions in the enlargement column generally don't coincide with the q and g in Equations (5.50) and (5.52). As mentioned, this is because of the necessary simplication when writing the enlargement ODEs. (These equations would occasionally be an unwieldy mess if this wasn't done.) After the integration, the original x, y variables are reintroduced, if necessary, thus yielding the final form for the solution.

Generic symbols, such as those in Table 5.2, are helpful for table-making but are of little use for problem-solving. In the latter situation, explicit ODEs are preferred, such as those in Table 5.1 or Equations (5.41) and (5.43). The generality of the group approach is evident in the enlargement column. For instance, the ODEs for item Ia1 involve three arbitrary functions: namely, M, q, and g. Any specific ODE that conforms to these equations is solvable. However, Table 5.4 is certainly not exhaustive, e.g., none of the symbols involve an error function. Also note that item Ia1 and the first enlargement of item Ibs1 are linear, and therefore Equation (5.53a) directly yields a solution. This compendium is a first attempt; hopefully, it will be improved upon and added to.

In practice, Equations (5.26) and (5.33) are written in terms of x, y, and y' to see if one of them matches the ODE to be solved. For instance, for item Ic3, Equation (5.33) yields

$$\frac{(1-m)y - nxy'}{x^m y^{n+1} y'} = g(y),$$

which can be written more conveniently as

$$\frac{x^m y'}{(1-m)y - nxy'} = g(y),$$

where the y^{n+1} factor in dY/dX is absorbed in the g function.

To further illustrate how the qs and gs are determined, the two enlargement ODEs for item Ic1 are evaluated. We start with

$$X = y, \quad Y = \int \frac{dx}{M(x)N(y)} = \frac{1}{N(y)} \int \frac{dx}{M(x)},$$

where N can be taken outside the integral, since $y = c$. The dY/dX derivative stems from

$$dX = dy$$

$$dY = -\frac{N'}{N^2} dy \int \frac{dx}{M} + \frac{dx}{MN} = -\frac{N'}{N} Y \, dy + \frac{dy}{MNy'}$$

$$= (1 - MN'Yy') \frac{dy}{MNy'},$$

while the first ODE in the enlargement column is based on Equation (5.50)

$$\frac{1 - MN'Yy'}{MNy'} + \frac{1}{N} \left(\int \frac{dx}{M} \right) q(y) = g(y).$$

Multiply by MN and rearrange, with the result

$$\frac{1}{y'} - MN'\frac{1}{N} \int \frac{dx}{M} + M \left(\int \frac{dx}{M} \right) q = MNg,$$

or

$$\frac{1}{y'} + M \left(\int \frac{dx}{M} \right) \left(q - \frac{N'}{N} \right) = M(Ng),$$

where $q - N'/N$ and Ng are replaced with $q(y)$ and $g(y)$, respectively. Equation (5.52) is used for the second ODE in the enlargement column, providing its simplified form is not redundant with previous ODEs. In this case, the second ODE is directly obtained from the X, Y, and dY/dX columns.

Type I in Table 5.3 is the simplest category, in which X equals x or y. For (a) through (d), we readily obtain:

(a) $Y = \int \dfrac{dx}{M(x)}, \quad \dfrac{dY}{dX} = \dfrac{1}{M\,y'},$

(b) $Y = \dfrac{x}{N(y)}, \quad \dfrac{dY}{dX} = \dfrac{1}{N\,y'} - \dfrac{x}{N^2}\dfrac{dN}{dy},$

(c) $Y = \dfrac{1}{N(y)}\int \dfrac{dx}{M(x)}, \quad \dfrac{dY}{dX} = \dfrac{1}{M\,N\,y'} - \dfrac{1}{N^2}\dfrac{dN}{dy}\int \dfrac{dx}{M},$

(d) $y = \int \dfrac{dx}{\xi(x,c)}, \quad \dfrac{dY}{dX} = \dfrac{1}{\xi\,y'} - \int_{y=c} \dfrac{\partial \xi}{\partial y}\dfrac{dx}{\xi^2}.$

The other cases are more involved. For instance, in type II, X equals

$$X = y - \int \frac{\eta(x)}{M(x)}\,dx$$

for (a), but changes to

$$X = \int N(y)dy - \int \eta(x)dx$$

for (b). Moreover, for (b), either $y = w(x,c)$ or $x = w(y,c)$ is required for Y. For example, in IIb1, the integrals

$$\int \frac{dx}{N(y)}, \quad \int \frac{N'dx}{N^2}$$

both require $y = w(x,c)$.

5.5   Examples

A variety of examples are discussed to illustrate the theory. We start with a simple one; namely, find the general solution of

$$y' = \frac{y - 3x\ln(x^2 + y^2)}{x + 3y\ln(x^2 + y^2)}. \tag{5.55}$$

This ODE satisfies rotation-group invariance, Equation (5.18c), with

$$g(x^2 + y^2) = 3\ln\left(x^2 + y^2\right) = 3\ln X.$$

Equation (5.34b) thus becomes

$$\frac{dY}{dX} = -\frac{1}{2X}\frac{y - xy'}{x + yy'} = -\frac{1}{2X}\frac{y - \left(\dfrac{xy - 3x^2 \ln X}{x + 3y \ln X}\right)}{x + \left(\dfrac{y^2 - 3xy \ln X}{x + 3y \ln X}\right)}$$

$$= -\frac{1}{2X}\frac{3(x^2 + y^2)\ln X}{(x^2 + y^2)} = -\frac{3}{2}\frac{\ln X}{X},$$

where X and Y are given by Equations (5.35). This equation is not only separable, but is readily integrated

$$\int dY = -\frac{3}{2}\int (\ln X)\frac{dX}{X} = -\frac{3}{2}\int \ln X \, d\ln X$$

with the result

$$Y = c - \frac{3}{4}(\ln X)^2,$$

or

$$\tan^{-1}\left(\frac{y}{x}\right) = c - \frac{3}{4}\left[\ln(x^2 + y^2)\right]^2. \tag{5.56}$$

With c as the constant of integration, this is the general solution of Equation (5.55).

The foregoing equations provide an opportunity to describe the theory from an alternate viewpoint. Invariance under a one-parameter transformation, Equations (2.3), means that Equation (5.1) transforms to

$$f(x_1, y_1, y_1') = 0,$$

where the functional dependence is the same as in Equation (5.1). For instance, Equation (5.55) is invariant under Equations (2.2) for the rotation group. For convenience, the inverse transformation

$$x = x_1 \cos \alpha - y_1 \sin \alpha$$
$$y = x_1 \sin \alpha + y_1 \cos \alpha$$

is substitute into Equation (5.55). After simplification, we obtain

$$y_1' = \frac{y_1 - 3x_1 \ln(x_1^2 + y_1^2)}{x_1 + 3y_1 \ln(x_1^2 + y_1^2)}$$

which is in accord with our invariance expectation.

The second example is based on the rotation group example in Section 4.1. We first determine the differential equation for a family of straight lines that are tangent to the unit circle shown in Figure 4.1. This is done by eliminating the constant a from Equation (4.13) by differentiation. We thereby obtain

$$a + (1 - a^2)^{1/2}y' = 0,$$

or

$$a = -\frac{y'}{[1 + (y')^2]^{1/2}},$$

which combines with Equation (4.13) to yield

$$y - xy' = [1 + (y')^2]^{1/2}. \tag{5.57a}$$

When this is solved for y', we have

$$y' = \frac{1}{1 - x^2}\left[-xy \pm (x^2 + y^2 - 1)^{1/2}\right]. \tag{5.57b}$$

Either of these equations is an ODE that is invariant under the rotation group. In fact, Equation (5.57b) is in accord with item 6 in Table 5.1 by setting

$$g = \pm(x^2 + y^2 - 1)^{-1/2}.$$

The solution of Equations (5.57) is a family of integral curves and these are straight lines furnished by Equation (4.13), with a as the constant of integration. (A second solution is the envelope of the family, which is the unit circle.) Thus, the path curves, which are circles, and integral curves, which are straight lines, are different. Outside of the unit circle, they form a nonorthogonal mesh in the x, y plane. Notice that a fixed point on a straight-line integral curve generates a spiral if the line does not slide along the unit circle. For a circular-arc path curve, the line must slide on the circle as it rotates.

In the next example, results are derived for item Ic5 in Table 5.4, starting with the symbol. In this case, we have

$$\xi = \exp(mx + ny), \quad \eta = 0,$$

from which we obtain

$$\frac{dx}{\exp(mx + ny)} = \frac{dy}{0}.$$

This results in

$$X = u = y = c,$$

and

$$Y = \int \frac{dx}{\xi} = \int \frac{dx}{e^{nc}e^{mx}} = e^{-nc}\int e^{-mx}\,dx = -\frac{1}{m}\exp[-(mx + ny)]$$

after replacing c with y. By differentiation

$$dX = \frac{dy}{dx}dx = y'\,dx$$

$$dY = \frac{1}{m}\exp[-(mx + ny)](mdx + ndy) = -Y(m + ny')dx,$$

and dY/dX is obtained as shown in Table 5.4.

The first enlargement in item Ic5 is based on Equation (5.50), written as

$$-\frac{m+ny'}{y'}Y + Yq(y) = g(y),$$

which becomes, after division by Y,

$$\frac{m}{y'} + n - q(y) = e^{mx}(-e^{ny}g),$$

or

$$\frac{1}{y'} + \left[\frac{n-q}{m}\right] = e^{mx}\left[-\frac{1}{m}e^{ny}g\right].$$

The bracketed terms are respectively replaced with $q(y)$ and $g(y)$, as shown in the table.

The second enlargement is based on Equation (5.52), which becomes

$$-\frac{y'}{(m+ny')\exp[-(mx+ny)]} + yq[\exp(mx+ny)] = g.$$

This can be revised to

$$\frac{y'}{m+ny'} + y\{-\exp[-(mx+ny)]q\} = \{-\exp[-(mx+ny)]g\}.$$

The two braced terms are respectively replaced with q and g as shown in the table. Thus, the q and g functions for the two Ic5 ODEs in the enlargement column are not only functionally different but have different arguments. Note that the exponential argument in the second enlargement can be replaced with $mx+ny$.

In the next example, the solution is found for

$$y' = \frac{1}{x^2 e^{-y^2} - xy}. \tag{5.58}$$

In view of the right side, the equation is written as

$$\frac{1}{y'} + xy = x^2 e^{-y^2},$$

which fits the first enlargement of item Ic2 in Table 5.4, with

$$q(y) = y, \quad g(y) = e^{-y^2}, \quad a = 2$$

and with $P(y)$ still undetermined. Recall that these g and q values may not coincide with those in Equation (5.50). Therefore, Table 5.4 is used to write Equation (5.50) as

$$\frac{(1-a)[1+xP(y)y']}{x^a B(y)y'} + \frac{x^{1-a}}{B}\widehat{q}(y) = \widehat{g}(y),$$

where the circumflex symbol is used to avoid confusion with the previous q and g functions. We solve for $1/y'$, with the result

$$\frac{1}{y'} + x\left[P + \frac{\hat{q}}{(1-a)B}\right] = x^a\left(\frac{B\hat{g}}{1-a}\right)$$

and thus obtain

$$q = P + \frac{\hat{q}}{(1-a)B} = y, \quad a = 2, \quad g = -B\hat{g} = e^{-y^2}.$$

We can choose $\hat{g}$ as

$$\hat{g} = -e^{-y^2} = -e^{-X^2},$$

which, in turn, leads to

$$B = 1, \quad P = 0, \quad \hat{q} = -y = -X,$$

where the introduction of X is convenient for the next step. The quadratures in Equation (5.53a) become:

$$\int \hat{q}dX = -\int XdX = -\frac{1}{2}X^2,$$

$$\int \hat{g}\exp\left(\int \hat{q}dX\right)dX = -\int e^{-X^2}e^{-X^2/2}dX = -\int e^{-3X^2/2}dX.$$

With the substitution

$$z = \left(\frac{3}{2}\right)^{1/2}X,$$

we obtain

$$\int \hat{g}\exp\left(\int \hat{q}dX\right)dX = -\left(\frac{2}{3}\right)^{1/2}\int e^{-z^2}dz = -\left(\frac{\pi}{6}\right)^{1/2}erf\left[\left(\frac{3}{2}\right)^{1/2}X\right],$$

where erf stands for the error function. Equation (5.53a) now becomes

$$Y = e^{X^2/2}\left\{c_1 - \left(\frac{\pi}{6}\right)^{1/2}erf\left[\left(\frac{3}{2}\right)^{1/2}X\right]\right\},$$

or, upon returning to the original variables,

$$\frac{1}{x} = e^{y^2/2}\left\{c_1 - \left(\frac{\pi}{6}\right)^{1/2}erf\left[\left(\frac{3}{2}\right)^{1/2}y\right]\right\}, \tag{5.59}$$

where, from Table 5.4,

$$Y = \frac{x^{1-a}}{B} = \frac{1}{x}.$$

Equation (5.59) is the general solution of Equation (5.58). It is worth noting that the Ic2 enlarged group is not the only one that can be applied to Equation (5.58) nor is it necessarily the simplest choice.

In the next example, the general solution is found for

$$\frac{xy'}{y + xy'} = r(xy)^s \ln y + j(xy)^k,\tag{5.60}$$

where j, k, r, and s are constants. From Table 5.4, observe that the first enlargement of item V2 fits, providing we choose

$$a = 1, \quad b = 1, \quad m = 0, \quad n = 1,$$

$$q(xy) = -r(xy)^s, \quad g(xy) = j(xy)^k.$$

With this choice, the canonical variables are

$$X = xy, \quad Y = \ln y,$$

and the first derivative is

$$\frac{dY}{dX} = \frac{1}{xy}\frac{xy'}{xy' + y} = \frac{1}{X}(rX^sY + jX^k).$$

This becomes Equation (5.50) by setting

$$\hat{q} = -rX^{s-1}, \quad \hat{g} = jX^{k-1}.$$

(Observe that $\hat{q}$ and $\hat{g}$ are functionally different from q and g, respectively.) The quadratures in Equation (5.53a) become

$$\int \hat{q}dX = -\frac{r}{s}X^s$$

$$\int \hat{g}\exp\left(\int \hat{q}dX\right) dX = j\int X^{k-1}\exp(-rX^s/s)dX.$$

Equation (5.53a) has the form

$$\ln y = \exp[r(xy)^s/s]\left[c_1 + j\int X^{k-1}\exp(-rX^s/s)dX\right],\tag{5.61}$$

where xy replaces X after the remaining quadrature is performed. Equation (5.61) is the desired solution; it can be verified by differentiation with respect to x. As a consequence of the integral, a dX/dx factor, which equals $y + xy'$, appears on the right side when this differentiation is performed.

In the next example, the enlargement equation for item Id2 is derived, and this result is used to solve

$$y' = \frac{1}{x^2y - xy^2}.\tag{5.62}$$

For this symbol, the canonical coordinates are provided by Equations (5.31). With Equation (5.30b) providing dY, we have

$$\frac{dY}{dX} = \frac{1}{Ay'} - \int \frac{\partial A}{\partial y} \frac{dx}{A^2} \tag{5.63}$$

with $y = c$ in the integrand. Thus, Equation (5.50) can be written as

$$\frac{1}{Ay'} - \int \frac{\partial A}{\partial y} \frac{dx}{A^2} + Yq(y) = g. \tag{5.64}$$

The X, Y interchange form is too messy to be listed in the table.

At first glance, it appears unlikely that Equation (5.62) would be invariant under this group. We nevertheless examine this possibility by first aligning the derivative terms in Equations (5.62) and (5.64)

$$\frac{1}{y'} + xy^2 = x^2 y \tag{5.65a}$$

$$\frac{1}{y'} - A \int \frac{\partial A}{\partial y} \frac{dx}{A^2} + AYq(y) = Ag(y). \tag{5.65b}$$

By comparing the two equations, we guess that A can be written as

$$A = (xy)^m, \tag{5.66}$$

where $m \neq 1$, since Equation (5.62) does not contain a logarithm factor. We thus obtain

$$Y = -\frac{1}{m-1} \frac{x}{(xy)^m}$$

$$\int \frac{\partial A}{\partial y} \frac{dx}{A^2} = \frac{m}{y^{m+1}} \int \frac{dx}{x^m} = -\frac{m}{m-1} \frac{1}{(xy)^m} \frac{x}{y}$$

and Equation (5.65b) reduces to

$$\frac{1}{y'} + x \left(\frac{m}{m-1} \frac{1}{y} - \frac{1}{m-1} q \right) = (xy)^m g.$$

By comparing this with Equation (5.65a), we readily obtain

$$m = 2, \quad q = \frac{2}{y} - y^2, \quad g = \frac{1}{y}.$$

The fact that q and g are functions only of y confirms this choice for A. We now have

$$Y = -\frac{1}{xy^2}, \quad q = \frac{2}{X} - X^2, \quad g = \frac{1}{X},$$

and

$$\int q\,dX = \ln X^2 - \frac{1}{3}X^3$$

$$e^{\int q\,dX} = X^2 e^{-X^3/3}$$

$$\int g e^{\int q\,dX}\,dX = \int e^{-X^3/3} X\,dX.$$

Equation (5.53a) thus becomes

$$Y = \frac{1}{X^2} e^{X^3/3}\left(-c_1 + \int e^{-X^3/3} X\,dX\right).$$

In terms of the original variables, we have

$$x = \frac{e^{-y^3/3}}{c_1 - \int e^{-y^3/3} y\,dy} \tag{5.67}$$

for the solution of Equation (5.62).

As a check, Equation (5.67) is differentiated with respect to x, with the result

$$1 = -\frac{y^2 e^{-y^3/3} y'}{c_1 - \int e^{-y^3/3} y\,dy} + \frac{e^{-y^3/3}(ye^{-y^3/3})y'}{[c_2 - \int e^{-y^3/3} y\,dy]^2}.$$

This can be simplified to

$$1 = -xy^2 y' + yx^2 y',$$

which agrees with Equation (5.62).

As a final example, the general solution of

$$y' = \frac{erf(y)}{\frac{2}{\pi^{1/2}}xe^{-y^2} - axy\,erf(y) + by(erf y)^2} \tag{5.68a}$$

is obtained, where a and b are constants. We first rewrite the equation as

$$\frac{1}{y'} + x\left(ay - \frac{2e^{-y^2}}{\pi^{1/2}erf(y)}\right) = by\,erf(y). \tag{5.68b}$$

With item Ib1, we have

$$q(y) = ay - \frac{2e^{-y^2}}{\pi^{1/2}erf(y)}, \quad g(y) = by\,erf(y)$$

and

$$X = y, \quad Y = \frac{x}{N(y)}, \quad \frac{dY}{dX} = \frac{N - xN'y'}{N^2 y'}.$$

Equation (5.50) yields

$$\frac{N - xN'y'}{N^2 y'} + \frac{x}{N}\widehat{q}(y) = \widehat{g}(y),$$

which is rewritten as

$$\frac{1}{y'} + x\left(\widehat{q} - \frac{N'}{N}\right) = N\widehat{g}.$$

A comparison with Equation (5.68b) produces

$$\widehat{q} = ay - \frac{2e^{-y^2}}{\pi^{1/2}erf(y)} + \frac{N'}{N}, \quad \widehat{g} = \frac{byerf(y)}{N}.$$

A simple result occurs if the nonunique value for N is chosen as

$$N = erf(y), \quad N' = \frac{2}{\pi^{1/2}}e^{-y^2}$$

so that

$$\widehat{q} = ay, \quad \widehat{g} = by.$$

Consequently, Y becomes $x/(erfy)$. Two of the integrals that appear in Equation (5.53a) are evaluated as follows:

$$\int qdX = \int\left(ay - \frac{N'(y)}{N(y)}\right)dy = \frac{1}{2}ay^2 - \ln(erfy)$$

$$\exp\left(\int qdX\right) = \frac{e^{ay^2/2}}{erfy}$$

$$\exp\left(-\int qdX\right) = (erfy)e^{-ay^2/2}.$$

The solution provided by Equation (5.53a) has the form

$$Y = (erfy)e^{-ay^2/2}\left(c_1 + b\int ye^{ay^2/2}dy\right),$$

or

$$x = \left(\frac{b}{a} + c_1 e^{-ay^2/2}\right)(erfy)^2,$$

where c_1 is an integration constant.

5.6 Problems

5.1. Determine Equation (5.7a) for each of the transformations in Problem 2.2.

5.2. Determine Equation (5.7a) for each of the transformations in Problem 2.4.

5.3. Determine Equation (5.7a) for each of the transformations in Problem 2.9.

5.4. Find the first-order ODE invariant under the group

$$Uf = \phi(x)f_x + e^{ay}f_y,$$

where a is a constant and ϕ is an arbitrary function of x.

5.5. Use the Riccati equation approach to derive an equation for $u_{Ry}^{(1)}$.

5.6. Find the canonical variables X and Y such that each ODE in Problem 5.1 has the form of Equation (5.26).

5.7. Repeat Problem 5.6 for the ODEs in Problem 5.2 using Equation (5.39) as the separable form.

5.8. Repeat Problem 5.6 for the ODEs in Problem 5.3.

5.9. Find the solution to the following ODEs:

$$\frac{xy' - y}{xy' + y} = x^{n-1}y^{n+1}, \tag{1}$$

$$y' = \frac{y}{x}\frac{1 + x^a y^{a+2}}{1 - x^a y^{a+2}}, \tag{2}$$

$$\ln\left(\frac{xy' - y}{xy' + y}\right) = \left\{\sin^{-1}\left[(xy)^{1/3}\right]\right\}^2 + \ln\left(\frac{y}{x}\right), \tag{3}$$

where n and a are constants. Use c for the constant of integration, and verify that your solution to the third ODE is correct.

5.10. Solve the ODEs:

$$(y')^2 + \frac{2y}{x}y' - 1 - 2y - y^2 + \left(\frac{y}{x}\right)^2 = 0 \tag{1}$$

$$(y')^2 - \frac{2y}{x}y' + \left(\frac{y}{x}\right)^2 = \frac{a}{xy}\ln\left(\frac{x}{y}\right). \tag{2}$$

5.11. Suppose that $u_i, 1, 2$, are the invariants of $U_i f = 0$. Show that

$$u = u_1 + \gamma u_2,$$

where γ is a nonzero constant, in general is not the invariant of

$$Uf = U_1 f + \gamma U_2 f.$$

5.12. Solve the ODE

$$y' = -\frac{x}{1-x^2}y - y^2 e^x.$$

5.13. Solve the ODE

$$xy^2 (y')^2 - y^3 y' + x = 0.$$

The final form of the solution should not involve a quadrature; simplify your results.

5.14. Find the general solution to

$$y' = \frac{y - 3x \ln(x^2 + y^2) + x(x^2 + y^2)^{1/2} \sin^{-1}[y(x^2 + y^2)^{-1/2}]}{x + 3y \ln(x^2 + y^2) - y(x^2 + y^2)^{1/2} \sin^{-1}[y(x^2 + y^2)^{-1/2}]}.$$

5.15. Find the symbol in Table 5.4 that represents the nonuniform magnification group

$$x_1 = \alpha^m x, \quad y_1 = \alpha^n y,$$

where $\alpha = 1$ for the identity transformation. Find the ODE invariant under this group and its enlargements.

5.16. Find the solution to

$$y' = \frac{3y^2}{y^2 + 2y - 4 \exp(3x - y)}.$$

5.17. Find the solution to

$$y' = -\frac{a}{x} + bx^e \exp(my),$$

where a, b, e, and m are constants, and

$$e \neq am - 1.$$

5.18. Find the solution to

$$\frac{y - ax^b y^n y'}{x^b y'} + \left(\frac{x^{1-b}}{1-b} - \frac{a}{n} y^n \right) \frac{1}{\ln y} = \ln y,$$

where a, b, and n are constants.

5.19. Find the solution to the type II equation

$$xy' = y + \frac{x^2}{1 + (y/x)^2 e^{-x}}.$$

5.20. Determine the simplified form of the first-order ODE invariant under the group

$$Uf = a\sin^2 x f_x + b\cos^2 x f_y,$$

where a and b are constants.

5.21. Find the solution to

$$\frac{1}{y'} + x(\ln x)y^3 = x.$$

5.22. Use Equation (5.47) and $u_{Rx}^{(1)}$ to derive Equation (5.50).

5.23. Prove that

$$u^{(1)} = \xi \frac{du}{dx}$$

when $\xi = \xi(x)$, and that

$$u^{(1)} = \eta \frac{du}{dy}$$

when $\eta = \eta(y)$.

Chapter 6

Higher-Order ODEs Invariant Under a One-Parameter Group

When dealing with an ODE of second or higher order, a first step toward obtaining a solution is to reduce the order by one. Such a reduction is warranted even if a further reduction in order cannot be found. The lower-order equation is more amenable to a numerical solution and to analysis, e.g., by phase-plane techniques.

The *one-parameter* theory of the last chapter is extended to higher-order equations. Briefly, we show that if an nth-order ODE, with $n \geq 2$, is invariant under a group, its order can be reduced by one. In other words, a third-order equation, e.g., can be reduced to a second-order ODE. A further reduction may or may not be possible. After the reduction, the resulting ODE is usually not separable and a theory for canonical coordinates does not exist. Of course, if the initial ODE is second order and it is reduced to a first-order equation, this equation can be treated by the methods of the previous chapter. In this context, the first-order ODE may be invariant under a different group from that used for the second-order equation. When this occurs, the original second-order ODE is invariant under two different groups, as discussed in the next chapter.

The analysis summarized above is provided in Section 6.1. Much of this section is also required for the theory of two-parameter groups in Chapter 7. Section 6.2 describes the theory for finding the groups that leave a second or higher order ODE invariant. The theory associated with a system of coupled, first-order ODEs is discussed in Section 6.3. Section 6.4 provides a compendium of second-order ODEs invariant under a one-parameter group, and represents an extension of Tables 5.3 and 5.4. The chapter concludes, in Section 6.5, with a series of examples. Several of these are from engineering and physics.

6.1   Invariant Equations

First Approach

A second-order ODE

$$f(x, y, y', y'') = 0 \tag{6.1}$$

is considered that is invariant under a one-parameter group. For this, it is necessary and sufficient that

$$U^{(2)} f = 0 \quad \text{whenever } f = 0, \tag{6.2}$$

where we require that $df \neq 0$. We need to find the general solution of

$$\xi f_x + \eta f_y + \eta' f_{y'} + \eta'' f_{y''} = 0 \tag{6.3}$$

that is provided by the characteristic equations

$$\frac{dx}{\xi} = \frac{dy}{\eta} = \frac{dy'}{\eta'} = \frac{dy''}{\eta''} . \tag{6.4}$$

The solution of the leftmost equation is $u = c$, where u is the invariant function of the group. As discussed in Section 5.1, there are two equivalent equations involving dy'. These are Equations (5.5) whose solution $u^{(1)}$ is the first differential invariant.

For dy'', there are three equivalent equations. The solution to any one of these, written as,

$$u^{(2)}(x, y, y', y'') = c_2 \tag{6.5}$$

is called the *second differential invariant*. As shown by Equation (4.23), η'' is linear with respect to y''. In this regard, the dy''/η'' in term in Equations (6.4) is simpler than the dy'/η' term. If the equation to be solved is

$$\frac{dx}{\xi(x, y)} = \frac{dy''}{\eta''(x, y, y', y'')} \tag{6.6}$$

then, if necessary, both y and y' need to be eliminated by means of u and $u^{(1)}$. This elimination may be difficult, and even when easy may result in Equation (6.6) being extremely complicated. In short, it may be more difficult to solve than the ODE whose solution we are trying to obtain. These remarks apply equally well to the other choices for the dy'' equation. Alternate approaches for finding $u^{(2)}$ are therefore developed shortly. The general form of a second-order ODE invariant under the group then is

$$G\left(u, u^{(1)}, u^{(2)}\right) = 0 \quad \text{or} \quad u^{(2)} = g\left(u, u^{(1)}\right). \tag{6.7}$$

Equivalence

As mentioned, there are three equivalent choices for $u^{(2)}$, depending on the choice for the dy'' equation. Suppose $u^{(2)}$ and $\tilde{u}^{(2)}$ stem from two different dy'' equations. These two differential invariants are equivalent to each other, as discussed in Section 5.1. This means they are functionally related by an equation of the form

$$\tilde{u}^{(2)} = F\left[u(x,y), u^{(1)}(x,y,y'), u^{(2)}(x,y,y',y'')\right]. \tag{6.8}$$

The basis of this relation stems from the equations for the invariants

$$Uu = 0, \quad U^{(1)}u^{(1)} = 0, \quad U^{(2)}u^{(2)} = 0. \tag{6.9}$$

Equation (6.8) is now verified by showing that it yields the invariance condition

$$U^{(2)}\tilde{u}^{(2)} = 0.$$

We start with Equation (6.8) and evaluate $U^{(2)}\tilde{u}^{(2)}$

$$U^{(2)}\tilde{u}^{(2)} = \xi \left(F_u u_x + F_{u^{(1)}}u_x^{(1)} + F_{u^{(2)}}u_x^{(2)}\right) + \eta \left(F_u u_y + F_{u^{(1)}}u_y^{(1)} + F_{u^{(2)}}u_y^{(2)}\right)$$
$$+ \eta' \left(F_{u^{(1)}}u_{y'}^{(1)} + F_{u^{(2)}}u_{y'}^{(2)}\right) + \eta'' F_{u^{(2)}}u_{y''}^{(2)},$$

where the chain rule is used for the differentials

$$F_x = F_u u_x + F_{u^{(1)}}u_x^{(1)} + F_{u^{(2)}}u_x^{(2)}$$
$$\vdots$$
$$F_{y''} = F_{u^{(2)}}u_{y''}^{(2)}.$$

After rearrangement and with the aid of Equations (6.9), we have

$$U^{(2)}\tilde{u}^{(2)} = F_u(\xi u_x + \eta u_y) + F_{u^{(1)}}\left(\xi u_x^{(1)} + \eta u_y^{(1)} + \eta' u_{y'}^{(1)}\right)$$
$$+ F_{u^{(2)}}\left(\xi u_x^{(2)} + \eta u_y^{(2)} + \eta' u_{y'}^{(2)} + \eta'' u_{y''}^{(2)}\right)$$
$$= F_u Uu + F_{u^{(1)}}U^{(1)}u^{(1)} + F_{u^{(2)}}U^{(2)}u^{(2)} = 0.$$

Thus, $\tilde{u}^{(2)}$ and $u^{(2)}$ are equivalent to each other because they satisfy the same invariance condition. The equivalence discussion is not limited to just the first and second differential invariants, but clearly holds for differential invariants of any order.

Discussion

The extension of the theory that generates the nth-order ODE, for $n \geq 3$, that is invariant under a one-parameter group is self-evident from Equations

(6.1)–(6.4), and need not be repeated. Whereas a first-order ODE admits for invariance anywhere from none to an infinite number of linearly independent groups, a second-order ODE admits from none to eight linearly independent groups. For instance, the ODE

$$y'' = xy + \tan y'$$

is not invariant under any infinitesimal transformation (Cohen, 1911), whereas

$$y'' = 0$$

is invariant under eight linearly independent infinitesimal transformations (Cohen, 1911). More generally, an nth-order ODE, $n \geq 3$, admits from none to $n+4$ linearly independent infinitesimal transformations under which it is invariant.

The linearly independent groups under which an ODE is invariant form a *basis set* that spans the space containing the groups under which the ODE is invariant. Thus, if an ODE is invariant under a group, this group can be written as a linear combination of the groups in the basis set. This important property will be utilized in Chapter 7.

Second Approach

Equation (6.6), or its two alternate forms, is usually difficult or impossible to analytically solve. Thus, a result of considerable utility is that the second differential invariant can be written as

$$u^{(2)} = \frac{du^{(1)}}{du} = \frac{du^{(1)}/dx}{du/dx} = \frac{u_x^{(1)} + u_y^{(1)}y' + u_{y'}^{(1)}y''}{u_x + u_y y'}, \tag{6.10}$$

where $u^{(2)}$ is equivalent, as we shall shortly prove, to any second differential invariant stemming from Equations (6.4). A second approach for obtaining the second-order ODE invariant under the group is then

$$G\left(u, u^{(1)}, \frac{du^{(1)}}{du}\right) = 0, \tag{6.11}$$

or

$$\frac{du^{(1)}}{du} = g\left(u, u^{(1)}\right). \tag{6.12}$$

With u and $u^{(1)}$ as new variables, we have reduced a second-order equation to first order. This is the reduction mentioned at the start of the chapter. It stems from the fact that the number of arguments in Equation (6.11) is one less than in Equation (6.1). Recall that u is usually readily obtained, while $u^{(1)}$, e.g., can be obtained as the solution of a Riccati equation.

The u and $u^{(1)}$ variables in the above equations are not canonical coordinates, since these equations are usually not separable. When $n \geq 2$, there is no general transformation that yields a separable form, such as

$$\frac{dy'}{dx} = g(x), \quad y'\frac{dy'}{dy} = g(y),$$

and canonical coordinates don't exist.

For notational simplicity, we introduce

$$X = u(x, y), \quad \tilde{Y} = u^{(1)}(x, y, y'), \tag{6.13}$$

where a tilde is used to avoid confusing it with the canonical coordinate Y in Chapter 5. Equation (6.12) thus becomes

$$\frac{d\tilde{Y}}{dX} = g(X, \tilde{Y}). \tag{6.14}$$

Although this equation may not be separable, suppose a solution is found that has the form

$$G(X, \tilde{Y}) = G\left[u(x, y), \quad u^{(1)}(x, y, y')\right] = a, \tag{6.15}$$

where a is an integration constant. This relation is just Equation (5.7a), since the integration constant can be absorbed into G. Consequently, we have a first-order ODE that is invariant under the *same group* that Equation (6.1) is invariant under. The above equation can be solved by the methods of Chapter 5, including the use of canonical coordinates. Thus, Equation (6.1) can be fully solved, in terms of quadratures, if it is twice invariant under the same group.

Proof that $u^{(2)}$ is a Second Differential Invariant

To prove that $u^{(2)}$, given by Equation (6.10), is a second differential invariant, we first evaluate

$$U^{(2)}\frac{du}{dx} = \xi\frac{\partial}{\partial x}\left(\frac{du}{dx}\right) + \eta\frac{\partial}{\partial y}\left(\frac{du}{dx}\right) + \eta'\frac{\partial}{\partial y'}\left(\frac{du}{dx}\right) + \eta''\frac{\partial}{\partial y''}\left(\frac{du}{dx}\right), \tag{6.16a}$$

where x, y, y', and y'' are the independent variables, and where

$$\frac{du}{dx} = u_x + u_y y'$$

$$\eta' = \frac{d\eta}{dx} - y'\frac{d\xi}{dx}. \tag{6.17}$$

Consequently, the partial derivatives become

$$\frac{\partial}{\partial x}\left(\frac{du}{dx}\right) = u_{xx} + u_{xy}y'$$

$$\frac{\partial}{\partial y}\left(\frac{du}{dx}\right) = u_{xy} + u_{yy}y'$$

$$\frac{\partial}{\partial y'}\left(\frac{du}{dx}\right) = u_y$$

$$\frac{\partial}{\partial y''}\left(\frac{du}{dx}\right) = 0$$

and the equation simplifies to

$$U^{(2)}\frac{du}{dx} = \xi\frac{du_x}{dx} + \eta\frac{du_y}{dx} + u_y\frac{d\eta}{dx} - u_y y'\frac{d\xi}{dx}. \tag{6.16b}$$

The invariance condition

$$Uu = \xi u_x + \eta u_y = 0$$

is differentiated with respect to x, to obtain

$$\xi\frac{du_x}{dx} + \eta\frac{du_y}{dx} + u_y\frac{d\eta}{dx} = -u_x\frac{d\xi}{dx}.$$

Equation (6.16b) now simplifies to

$$U^{(2)}\frac{du}{dx} = -u_x\frac{d\xi}{dx} - u_y y'\frac{d\xi}{dx} = -\frac{du}{dx}\frac{d\xi}{dx}. \tag{6.18}$$

In a similar manner, we evaluate

$$U^{(2)}\frac{du^{(1)}}{dx} = \xi\frac{\partial}{\partial x}\frac{du^{(1)}}{dx} + \eta\frac{\partial}{\partial y}\frac{du^{(1)}}{dx} + \eta'\frac{\partial}{\partial y'}\frac{du^{(1)}}{dx} + \eta''\frac{\partial}{\partial y''}\frac{du^{(1)}}{dx}, \tag{6.19}$$

where

$$\frac{du^{(1)}}{dx} = u_x^{(1)}(x,y,y') + u_y^{(1)}(x,y,y')y' + u_{y'}^{(1)}(x,y,y')y''$$

$$\eta'' = \frac{d\eta'}{dx} - y''\frac{d\xi}{dx}. \tag{6.20}$$

The partial derivatives become

$$\frac{\partial}{\partial x}\frac{du^{(1)}}{dx} = u_{xx}^{(1)} + u_{xy}^{(1)}y' + u_{xy'}^{(1)}y''$$

$$\frac{\partial}{\partial y}\frac{du^{(1)}}{dx} = u_{xy}^{(1)} + u_{yy}^{(1)}y'' + u_{yy'}^{(1)}y''$$

$$\frac{\partial}{\partial y'}\frac{du^{(1)}}{dx} = u_{xy'}^{(1)} + u_{yy'}^{(1)}y' + u_{y'y'}^{(1)}y'' + u_y^{(1)}$$

$$\frac{\partial}{\partial y''}\frac{du^{(1)}}{dx} = u_{y'}^{(1)}.$$

Note that these derivative computations do not necessarily commute, e.g.,

$$\frac{\partial}{\partial y'}\frac{du^{(1)}}{dx} \neq \frac{du^{(1)}_{y'}}{dx}.$$

With the above relations and Equations (6.20), we have

$$U^{(2)}\frac{du^{(1)}}{dx} = \xi\frac{du^{(1)}_x}{dx} + \eta\frac{du^{(1)}_y}{dx} + \eta'\frac{du^{(1)}_{y'}}{dx} + u^{(1)}_{y'}\frac{d\eta'}{dx} + \eta'u^{(1)}_y - u^{(1)}_{y'}y''\frac{d\xi}{dx}. \quad (6.21)$$

The invariance condition

$$U^{(1)}u^{(1)} = \xi u^{(1)}_x + \eta u^{(1)}_y + \eta'u^{(1)}_{y'} = 0$$

is differentiated with respect to x to obtain

$$\xi\frac{du^{(1)}_x}{dx} + \eta\frac{du^{(1)}_y}{dx} + \eta'\frac{du^{(1)}_{y'}}{dx} + u^{(1)}_{y'}\frac{d\eta'}{dx} = -u^{(1)}_x\frac{d\xi}{dx} - u^{(1)}_y\frac{d\eta}{dx}.$$

Consequently, Equation (6.21) simplifies to

$$U^{(2)}\frac{du^{(1)}}{dx} = -\left(u^{(1)}_x + u^{(1)}_y y' + u^{(1)}_{y'}y''\right)\frac{d\xi}{dx} = -\frac{du^{(1)}}{dx}\frac{d\xi}{dx}, \quad (6.22)$$

which is a straightforward generalization of Equation (6.18).

We now show that $u^{(2)}$, given by Equation (6.10), is a second differential invariant by showing that

$$U^{(2)}u^{(2)} = U^{(2)}\left[\frac{du^{(1)}}{dx}\left(\frac{du}{dx}\right)^{-1}\right]$$

is zero. We obtain

$$U^{(2)}u^{(2)} = \frac{\dfrac{du}{dx}U^{(2)}\dfrac{du^{(1)}}{dx} - \dfrac{du^{(1)}}{dx}U^{(2)}\dfrac{du}{dx}}{\left(\dfrac{du}{dx}\right)^2} = \frac{-\dfrac{du}{dx}\dfrac{du^{(1)}}{dx}\dfrac{d\xi}{dx} + \dfrac{du^{(1)}}{dx}\dfrac{du}{dx}\dfrac{d\xi}{dx}}{\left(\dfrac{du}{dx}\right)^2} = 0,$$

where Equations (6.18) and (6.22) are used. One can show that the higher-order differential invariants are furnished by (see Problem 6.16)

$$u^{(3)} = \frac{d^2u^{(1)}}{du^2} = \frac{du^{(2)}}{du}, \quad u^{(4)} = \frac{d^3u^{(1)}}{du^3}, \cdots.$$

Third Approach

There is a particularly useful third approach (Ovsiannikov, 1983) that generalizes Equation (5.31). As shown in Section 5.2, dY/dX is equivalent to $u^{(1)}$. Hence, from Equation (6.11), we have

$$G\left(X, \frac{dY}{dX}, \frac{d^2Y}{dX^2}\right) = 0 \ \text{ or } \ \frac{d^2Y}{dX^2} = g\left(X, \frac{dY}{dX}\right), \tag{6.23}$$

where $X = u$ and Y is given by either of Equations (5.29). More generally, we can write

$$G\left(X, Y', Y'', \dots, Y^{(n)}\right) = 0 \tag{6.24}$$

where

$$Y^{(n)} = \frac{d^n Y}{dX^n}$$

for the nth-order ODE invariant under the group. Observe that Y itself is not one of the arguments of the above equations. Consequently, the order of these equations can be reduced by one by simply replacing dY/dX with a new dependent variable.

6.2 Finding the Groups

An alternative approach for second- or higher-order ODEs is to find the groups under which a given ODE is invariant. This approach was discussed for a first-order ODE in Section 5.4, where it was dismissed because there was only one equation for determining ξ and η.

As we shall soon see, the same approach for a second-order ODE yields a system of several linear, second-order PDEs. This system contains more equations than unknowns, i.e., it is *overdetermined*. In contrast to an underdetermined system, a *determinant* system, where the number of unknowns and equations are equal, or an overdetermined system is amenable to analysis and, on occasion, may lead to useful results.

For third- and higher-order ODEs, this discussion generalizes in a straightforward manner. For example, a third-order ODE generates an overdetermined system of linear, third-order PDEs. What is not straightforward is the monumental effort needed to obtain and to solve the very large resulting system of PDEs. This task is best relegated to a computer (Schwarz, 1988).

We begin by considering the invariance condition when f is given by

$$f = g(x, y, y') - y'' = 0. \tag{6.25}$$

The second-order ODE counterpart of Equation (5.54) stems from

$$\xi f_x + \eta f_y + \eta' f_{y'} + \eta'' f_{y''} = 0.$$

With the elimination of f, η', η'', and y'', we have

$$g_x \xi + g_y \eta + g_{y'} \left[\eta_x + (\eta_y - \xi_x)y' - \xi_y(y')^2 \right] - \eta_{xx} - (2\eta_{xy} - \xi_{xx})y'$$
$$- (\eta_{yy} - 2\xi_{xy})(y')^2 + \xi_{yy}(y')^3 - g(\eta_y - 2\xi_x - 3\xi_y y') = 0. \qquad (6.26)$$

In this relation, ξ and η are unknown functions that depend only on x and y. The equation therefore can be *decomposed* into a sequence of equations wherein the coefficients of the various powers of y' are set equal to zero. This step, of course, requires knowing the dependence of g on y'.

For purposes of illustrating the concept, a polynomial form with respect to y' is assumed

$$g = \sum_{i=0}^{n} g_i(x, y)(y')^i = g_0 + g_1 y' + \cdots + g_n(y')^n. \qquad (6.27)$$

This relation is substituted into Equation (6.26). The above sorting process for the respective coefficients of $(y')^m$, $m = 0, \ldots, 3, \ldots, n+1$, then yields:

$$g_{0,x}\xi + g_{0,y}\eta + g_1\eta_x - \eta_{xx} + g_0(2\xi_x - \eta_y) = 0 \qquad (6.28\text{a})$$
$$g_{1,x}\xi + g_{1,y}\eta + 2g_2\eta_x + g_1\xi_x + \xi_{xx} - 2\eta_{xy} + 3g_0\xi_y = 0 \qquad (6.28\text{b})$$
$$g_{2,x}\xi + g_{2,y}\eta + 3g_3\eta_x + 2\xi_{xy} - \eta_{yy} + 2g_1\xi_y + g_2\eta_y = 0 \qquad (6.28\text{c})$$
$$g_{3,x}\xi + g_{3,y}\eta + 4g_4\eta_x + \xi_{yy} + g_2\xi_y + g_3(2\eta_y - \xi_x) = 0 \qquad (6.28\text{d})$$
$$g_{4,x}\xi + g_{4,y}\eta + 5g_5\eta_x + g_4(3\eta_y - 2\xi_x) = 0 \qquad (6.28\text{e})$$

$$\vdots$$

With $g_n \neq 0$, the final equation in the sequence depends on the value of n. It is given by:

$$\xi_{yy} = 0, \quad n = 0, 1 \qquad (6.28\text{f})$$
$$\xi_{yy} + g_2\xi_y = 0, \quad n = 2 \qquad (6.28\text{g})$$
$$g_{3x}\xi + g_{3,y}\eta + g_3(2\eta_y - \xi_x) + g_2\xi_y + \xi_{yy} = 0, \quad n = 3 \qquad (6.28\text{h})$$
$$\xi_y = 0, \quad n \geq 4. \qquad (6.28\text{i})$$

Equations (6.28), as does Equation (6.26), always admit the trivial solution

$$\xi \equiv 0, \ \eta \equiv 0.$$

This solution, however, does not represent a one-parameter group and is of no interest.

As an example, suppose we have just one term

$$g = g_0(x, y).$$

If g_0 only depends on x or on y, then Equation (6.25) is directly separable and can be integrated twice. Consequently, g_0 is assumed to depend on both x and

y. Equations (6.28) readily reduce to

$$g_{0,x}\xi + g_{0,y}\eta + g_0(2\xi_x - \eta_y) - \eta_{xx} = 0 \tag{6.29a}$$

$$\xi_{xx} - 2\eta_{xy} + 3g_0\xi_y = 0 \tag{6.29b}$$

$$2\xi_{xy} - \eta_{yy} = 0 \tag{6.29c}$$

$$\xi_{yy} = 0. \tag{6.29d}$$

From Equations (6.29c) and (6.29d), we deduce

$$\xi = a_1(x)y + a_2(x),$$
$$\eta = a_1'y^2 + a_3(x)y + a_4(x), \tag{6.30}$$

where the a_i are arbitrary functions of x. Equations (6.29b) and (6.30) then yield

$$g_0 = \frac{3a_1''y + 2a_3' - a_2''}{3a_1}. \tag{6.31}$$

This relation is very restrictive; g_0 is linear with respect to y, unless $a_1'' = 0$ and g_0 only depends on x. If we now substitute Equations (6.30) and (6.31) into Equation (6.29a) and sort on y, we obtain two complicated, coupled relations involving the a_i. These are nonlinear differential equations whose highest-order derivatives are a_1''', a_2''', a_3'', and a_4''. These two relations are likely to be more complicated than the particular ODE we are trying to solve.

The above decomposition procedure can be used to determine the number of linearly independent groups that a second- or higher-order ODE is invariant under. In particular, it can be used to demonstrate that a given ODE may not be invariant under any group. This becomes apparent when the only solution to the overdetermined system is for ξ and η to be identically zero. This occurs for the equation that has a $\tan(y')$ term in Section 6.1 and in Problem 6.23. The procedure, moreover, may also yield a system of ODEs that are as complicated or even more complicated than the equation one is trying to solve, e.g., see Problem 6.22. While this technique is not emphasized for solving ODEs, it dominates the invariance analysis when dealing with a PDE or a system of PDEs.

6.3 System of First-Order ODEs

Transformation to Two First-Order ODEs

Any second-order ODE, such as Equation (6.1), can be written as a system of two, first-order ODEs by introducing the new dependent variables

$$y_1 = y, \quad y_2 = y'. \tag{6.32}$$

The new system of equations becomes

$$f_1 = f(x, y_1, y_2, y_2') = 0 \tag{6.33a}$$
$$f_2 = y_2 - y_1' = 0. \tag{6.33b}$$

The concept of invariance under a one-parameter group for a system of equations is readily obtained from Equations (4.7). If the system consists of several first-order ODEs, we require

$$U^{(1)} f_k = 0, \quad k = 1, 2, \ldots \tag{6.34}$$

for the system to be invariant under a one-parameter group. For a system of two ODEs, this condition is written as

$$U^{(1)} f_k = \xi f_{k,x} + \eta_1 f_{k,y_1} + \eta_2 f_{k,y_2} + \eta_1' f_{k,y_1'} + \eta_2' f_{k,y_2'} = 0, \quad k = 1, 2$$

which simplifies to

$$\xi f_{1,x} + \eta_1 f_{1,y_1} + \eta_2 f_{1,y_2} + \eta_2' f_{1,y_2} = 0 \tag{6.35a}$$
$$\eta_2 - \eta_1' = 0 \tag{6.35b}$$

for Equations (6.33). By comparing these equations with Equation (6.3), the correspondence is obtained

$$x \rightarrow x$$
$$y \rightarrow y_1$$
$$y' \rightarrow y_2$$
$$y'' \rightarrow y_2'$$
$$\xi \rightarrow \xi(x, y_1)$$
$$\eta \rightarrow \eta_1(x, y_1)$$
$$\eta' \rightarrow \eta_2(x, y_1, y_2) = \eta_1'$$
$$\eta'' \rightarrow \eta_2'(x, y_1, y_2, y_2'), \tag{6.36}$$

where the quantities on the right-hand side are associated with the system of first-order ODEs. Observe that η' and η'' are known functions [see Equations (4.18b) and (4.23)], e.g.,

$$\eta_2 = \eta_1' = \eta_{1,x} + (\eta_{1,y_1} - \xi_x) y_2 - \xi_{y_1} y_2^2. \tag{6.37}$$

This method for transforming the invariance condition of a second-order ODE into that for two, first-order ODEs is extended in a straightforward manner to still higher-order ODEs.

If the group is known, then determining the general form for f_1, starting from Equation (6.35a), is entirely comparable to using the procedures in Section 6.1.

Alternatively, if f_1 is known, we may decide to seek the groups leading to invariance. The symbol for these groups has ξ, η_1, and η_2 as coefficients. The η_2

coefficient, however, is provided by the above equation. This leaves only $\xi(x, y_1)$ and $\eta_1(x, y_1)$, which are determined by Equation (6.35a). The procedure in Section 6.2 can then be used with an overdetermined system to find ξ and η_1, as will be illustrated in Section 6.5.

System of First-Order ODEs

Let us consider the system

$$f_k = g_k(x, y_1, \ldots, y_n) - y_k' = 0, \quad k = 1, \ldots, n \tag{6.38}$$

of first-order ODEs, where we have assumed the derivatives can be solved for explicitly. Such a system is often encountered in science and engineering. For example, equations of this form represent the evolution with time of a chemically reactive system or the population of several biological species that are in competition with each other. We further assume these equations are invariant under a one-parameter group whose symbol is

$$U f_k = \xi f_{k,x} + \sum_{j=1}^{n} \eta_j f_{k,y_j}, \quad k = 1, \ldots, n \tag{6.39}$$

Thus, each y_j is associated with an η_j.

The first extension of the symbol is given by

$$U^{(1)} f_k = U f_k + \sum_{j=1}^{n} \eta_j' f_{k,y_j'}, \quad k = 1, \ldots, n \tag{6.40}$$

and invariance requires

$$U^{(1)} f_k = 0, \quad k = 1, \ldots, n \tag{6.41}$$

whenever Equation (6.38) is satisfied. Here, η_k' is defined by

$$\eta_k' = \eta_{k,x} - \xi_x y_k' + \sum_{j=1}^{n} y_j' (\eta_{k,y_j} - y_k' \xi_{y_j}) \tag{6.42}$$

which is a straightforward generalization of Equation (4.14).

We differentiate Equation (6.38) with respect to y_j and utilize Equations (6.39) and (6.40), with the result

$$U^{(1)} f_k = g_{k,x} \xi + \sum_{j=1}^{n} g_{k,y_j} \eta_j - \eta_k', \quad k = 1, \ldots, n \tag{6.43}$$

This relation is combined with Equations (6.41) and (6.42) to obtain

$$g_{k,x} \xi - \eta_{k,x} + \xi_x y_k' + \sum_{j=1}^{n} (g_{k,y_j} \eta_j - y_j' \eta_{k,y_j} + y_j' y_k' \xi_{y_j}) = 0, \quad k = 1, \ldots, n.$$

Equation (6.38) is now used to eliminate the first derivatives, which yields the final result

$$g_{k,x}\xi + g_k\xi_x - \eta_{k,x}$$

$$+ \sum_{j=1}^{n}(g_{k,y_j}\eta_j - g_j\eta_{k,y_j} + g_jg_k\xi_{y_j}) = 0, \quad k = 1,\ldots,n, \tag{6.44}$$

which is a generalization of Equation (5.54). After the g_k and its derivatives are replaced, the unknowns are ξ and η_1 and the equations are overdetermined when $n \geq 2$.

A numerical solution of a system of first-order ODEs, such as Equation (6.38), can be obtained by utilizing a standard integration procedure, such as Runge-Kutta. Alternatively, a Lie series expansion for each ODE can be used. This method is discussed and illustrated, e.g., in Rodriguez Azara (1992) and in Hill (1982).

An analytical approach for a system of first-order ODEs would start with a known symbol and the η_k' are then determined using Equation (6.42). The characteristic equations are given by

$$\frac{dx}{\xi} = \frac{dy_1}{\eta_1} = \cdots = \frac{dy_n}{\eta_n} = \frac{dy_1'}{\eta_1'} = \cdots = \frac{dy_n'}{\eta_n'}.$$

The first n equations on the left may require a simultaneous solution, since ξ and η_j are functions of $x, y_1, \ldots, y_n$. In principle, the solution of these equations can be written as

$$u_k(x, y_1, \ldots, y_n) = c_k, \quad k = 1,\ldots,n,$$

where the c_k are integration constants.

We next solve the system

$$\frac{dx}{\xi} = \frac{dy_k'}{\eta_k'} \quad \text{or} \quad \frac{dy_k}{\eta_k} = \frac{dy_k'}{\eta_k'}, \quad k = 1,\ldots,n,$$

where these relations may also require a simultaneous solution, since the η_k' may depend on $y_1', \ldots, y_n'$ as well as $x, y_1, \ldots, y_n$. Again, in principle, the solution can be written as

$$u_k^{(1)}(x, y_1, \ldots, y_n, y_1', \ldots, y_n') = c_k^{(1)}, \quad k = 1,\ldots,n.$$

The system of ODEs invariant under the group then is

$$G_k\left(u_k, u_k^{(1)}\right) = 0, \quad k = 1,\ldots,n.$$

6.4   Compendium

Table 6.1

Table 6.1 provides a collection of second-order ODEs invariant under a group. The table is an extension of Tables 5.3 and 5.4; thus, the item column coincides with that in Table 5.4. All second derivatives in the table are obtained from

$$Y'' = \frac{d^2 Y}{dX^2} = \frac{dY'}{dx}\frac{dx}{dX} = \frac{Y'_x + Y'_y y' + Y'_{y'} y''}{X_x + X_y y'}, \tag{6.45}$$

where X and Y' are provided by Table 5.4. The ODE in the rightmost column is the simplest form obtainable from Equation (6.23), where g is an arbitrary function. The order of the transformed ODE is then reduced by one and the resulting first-order equation is analytically solved, if possible.

Not every item present in Table 5.4 is continued in Table 6.1. In some cases, the complexity of Y'' did not warrant its inclusion. The restrictions that occasionally appear in the symbol column of Table 5.4, of course, still apply. Examination of items Ic1 and V1 in Table 6.1 illustrate the complexity that may result from the use of a general form for the symbol. As a consequence, useful results are typically based on specific functional forms for ξ and η.

Substitution Principle

The substitution principle also applies, but with the addition of

$$y'' \rightleftharpoons -\frac{y''}{(y')^3} \tag{6.46}$$

to Equations (5.36). The manner of application is readily deduced, e.g., by comparing the first two items in Table 6.1.

Enlargement Procedure

The enlargement procedure is extended by differentiating $\widehat{Y}$ in Equations (5.49) twice, with the result

$$\frac{d\widehat{Y}}{dX} = \frac{1}{p}\frac{dY}{dX} - \frac{Y}{p^2}\frac{dp}{dX}$$

$$\frac{d^2\widehat{Y}}{dX^2} = \frac{1}{p}\frac{d^2 Y}{dX^2} - \frac{2}{p^2}\frac{dp}{dX}\frac{dY}{dX} - \frac{d}{dX}\left(\frac{1}{p^2}\frac{dp}{dX}\right)Y.$$

Equation (6.23), with $\widehat{Y}$ replacing Y, can be written as

$$\frac{1}{p}\frac{d^2 Y}{dX^2} - \frac{2}{p^2}\frac{dp}{dX}\frac{dY}{dX} - \frac{d}{dX}\left(\frac{1}{p^2}\frac{dp}{dX}\right)Y = g\left(X, \frac{1}{p}\frac{dY}{dX} - \frac{1}{p^2}\frac{dp}{dX}Y\right).$$

After simplification, we have

$$\frac{d^2Y}{dX^2} + 2q\frac{dY}{dX} + p\frac{d}{dX}\left(\frac{q}{p}\right)Y = g\left(X, \frac{dY}{dX} + Yq\right), \tag{6.47}$$

where $p = p(X)$ and

$$q(X) = -\frac{1}{p}\frac{dp}{dX}. \tag{6.48}$$

In Equation (6.47), g and p (or q) are arbitrary functions, where p and q are related by Equation (6.48). While the left side of Equation (6.47) is linear, the right side is generally nonlinear. In contrast to Equation (5.50), a general solution for Equation (6.47) is not possible because of the second argument in g. By introducing Equations (5.49), Equation (6.47) reverts to

$$\frac{d^2\widehat{Y}}{dX^2} = g\left(X, \frac{d\widehat{Y}}{dX}\right), \tag{6.49}$$

as expected. By setting $d\widehat{Y}/dX$ equal to a new dependent variable, a first-order equation is thereby obtained.

Once Equation (6.47) is found, X, Y, Y', and Y'' are replaced using Tables 5.4 and 6.1. The ODE is then simplified, with the result being a general form for a second-order ODE that is invariant under the enlarged group.

6.5 Examples

We determine $U^{(2)}$ and the second-order ODE invariant under the rotation group. This is the first of a series of examples involving this group. We consider that Uf, η', u, and $u^{(1)}$ are known; they may be found in Appendix B. From Equation (4.22), we have

$$\eta'' = \frac{d\eta'}{dx} - y''\frac{d\xi}{dx} = 2y'y'' - y''(-y') = 3y'y''$$

and Equations (6.4) become

$$-\frac{dx}{y} = \frac{dy}{x} = \frac{dy'}{1+(y')^2} = \frac{dy''}{3y'y''}.$$

To determine the second differential invariant, we use

$$\frac{dy''}{y''} = \frac{3y'}{1+(y')^2}dy' = \frac{3}{2}\frac{d[1+(y')^2]}{1+(y')^2}.$$

This yields

$$y'' = c_2\left[1+(y')^2\right]^{3/2},$$

or

$$u^{(2)} = \frac{y''}{[1 + (y')^2]^{3/2}} = c_2$$

for this invariant. The equation invariant under rotation thus is

$$G\left(x^2 + y^2, \frac{y - xy'}{x + yy'}, \frac{y''}{[1 + (y')^2]^{3/2}}\right) = 0, \tag{6.50a}$$

or

$$y'' = [1 + (y')^2]^{3/2} g\left(x^2 + y^2, \frac{y - xy'}{x + yy'}\right). \tag{6.50b}$$

The three rotational group invariants have the following meaning:

u = square of the radius vector to a point on the integral curve

$u^{(1)}$ = tangent of the angle between the radius vector and the tangent to the integral curve

$u^{(2)}$ = curvature of the integral curve

In the next example, we first derive the simplified ODE shown in Table 6.1 for item IIb3. The symbol for this item is a slight generalization of the rotation group symbol. We therefore show how the item IIb3 ODE in Table 6.1 reduces to Equation (6.50b).

The item IIb3 ODE, which is based on Equation (6.23), is written as

$$(bx^2 + ay^2)^2 y'' - 2(xy' - y)(ayy' + bx)^2 - (bx^2 + ay^2)(xy' - y)[a(y')^2 + b]$$
$$= \frac{4}{(ab)^{1/2}}(bx^2 + ay^2)^2(ayy' + bx)^3 g\left[bx^2 + ay^2, \frac{(ab)^{1/2}(y - xy')}{2(bx^2 + ay^2)(ayy' + bx)}\right].$$

The coefficient of g simplifies to $(ayy' + bx)^3$, while the second argument of g simplifies to

$$\frac{y - xy'}{ayy' + bx}.$$

The right side of the equation thereby becomes

$$(xy' - y)^3 \left(\frac{ayy' + bx}{xy' - y}\right)^3 g = (xy' - y)^3 g\left(bx^2 + ay^2, \frac{y - xy'}{ayy' + bx}\right).$$

The resulting equation is divided by the coefficient of y'' with the result

$$y'' - 2\frac{(xy' - y)^3}{(bx^2 + ay^2)^2}\left(\frac{ayy' + bx}{xy' - y}\right)^2 - \frac{(xy' - y)[a(y')^2 + b]}{(bx^2 + ay^2)} = (xy' - y)^3 g,$$

where, on the right side, the previous coefficient of y'' is absorbed into g. Similarly, the second term on the left side is also absorbed into $(xy' - y)^3 g$. After multiplication by $(bx^2 + ay^2)$, we obtain the ODE shown in Table 6.1.

To derive Equation (6.50b), set $a = b = 1$ in the item IIb3 ODE, with the result

$$\left(x^2 + y^2\right) y'' + \left(y - xy'\right)\left[1 + (y')^2\right] = \left(xy' - y\right)^3 g\left(x^2 + y^2, \frac{y - xy'}{x + yy'}\right). \quad (6.51)$$

We solve Equation (5.17) for y', to obtain

$$y' = \frac{y - xu^{(1)}}{x + yu^{(1)}}.$$

This result is used to evaluate

$$y - xy' = \frac{uu^{(1)}}{x + yu^{(1)}}$$

and

$$1 + (y')^2 = \frac{u[1 + (u^{(1)})^2]}{(x + yu^{(1)})^2}, \quad (6.52)$$

where u equals $x^2 + y^2$. Equation (6.51) can now be written as

$$y'' + \frac{uu^{(1)}[1 + (u^{(1)})^2]}{(x + yu^{(1)})^3} = -\frac{u^2(u^{(1)})^3}{(x + yu^{(1)})^3} g\left(u, u^{(1)}\right)$$

and the second term on the left, as well as $-u^2(u^{(1)})^3$, can be absorbed into the right-hand side. The equation, therefore, becomes

$$y'' = \frac{g}{(x + yu^{(1)})^3}.$$

However, Equation (6.52) yields

$$\frac{1}{(x + yu^{(1)})^3} = \frac{[1 + (y')^2]^{3/2}}{\{u[1 + (u^{(1)})^2]\}^{3/2}}$$

for the coefficient of g. With this substitution, Equation (6.50b) is finally obtained.

In the next example, a first-order ODE, Equation (6.14), is derived for the rotation group by introducing Equations (6.13):

$$X = u = x^2 + y^2$$
$$dX = 2dx(x + yy')$$
$$\tilde{Y} = u^{(1)} = \frac{y - xy'}{x + yy'} \quad (6.53)$$
$$d\tilde{Y} = dx\left(u_x^{(1)} + u_y^{(1)}y' + u_{y'}^{(1)}y''\right)$$
$$= -\frac{dx}{(x + yy')^2}\left\{\left[1 + (y')^2\right](y - xy') + (x^2 + y^2) y''\right\}.$$

We thus obtain

$$\frac{d\widetilde{Y}}{dX} = -\frac{1}{2\left(x + yy'\right)^3}\left\{\left[1 + \left(y'\right)^2\right]\left(y - xy'\right) + \left(x^2 + y^2\right)y''\right\}.$$

But the identity

$$\left(x^2 + y^2\right)\left[1 + \left(y'\right)^2\right] = \left(x + yy'\right)^2 + \left(y - xy'\right)^2$$

yields

$$\frac{d\widetilde{Y}}{dX} = -\frac{1}{2\left(x^2 + y^2\right)}\left[\frac{y - xy'}{x + yy'} + \left(\frac{y - xy'}{x + yy'}\right)^3 + \frac{\left(x^2 + y^2\right)^2}{\left(x + yy'\right)^3}y''\right]$$

$$= -\frac{1}{2X}\left(\widetilde{Y} + \widetilde{Y}^3 + \frac{\left(x^2 + y^2\right)^{1/2}}{\left(x + yy'\right)^3}\frac{y''}{[1 + (y')^2]^{3/2}}\{(x^2 + y^2)^{3/2}[1 + (y')^2]^{3/2}\}\right)$$

$$= -\frac{1}{2X}\left(\widetilde{Y} + \widetilde{Y}^3 + X^{1/2}(1 + \widetilde{Y}^2)^{3/2}\left\{\frac{y''}{[1 + (y')^2]^{3/2}}\right\}\right).$$

Observe that the braced factor containing y'' equals $u^{(2)}$ of the rotation group. By means of Equations (6.12) and (6.13), this factor can be replaced with

$$\frac{y''}{[1 + (y')^2]^{3/2}} = u^{(2)} = \frac{du^{(1)}}{du} = g(X, \widetilde{Y}).$$

Hence, Equation (6.14) becomes a first-order ODE

$$\frac{d\widetilde{Y}}{dX} = -\frac{1 + \widetilde{Y}^2}{2X}\left\{\widetilde{Y} + \left[X(1 + \widetilde{Y}^2)\right]^{1/2}g(X, \widetilde{Y})\right\} \tag{6.54}$$

where g is an arbitrary function of its arguments.

In general, the above ODE is not separable, since g is arbitrary. If it can be integrated, a first-order ODE is obtained by replacing X and $\widetilde{Y}$ with the original variables. However, this ODE is invariant under rotation, and the substitution of Equations (5.35) would put it into a separable form.

In the next example, suppose the ODE

$$\left(x^2 + y^2\right)y'' - \left[1 + \left(y'\right)^2\right]\left(y - xy'\right) = 0 \tag{6.55}$$

is invariant under rotation. We, therefore, rewrite it as

$$\frac{y''}{[1 + (y')^2]^{3/2}} = \frac{y - xy'}{(x^2 + y^2)[1 + (y')^2]^{1/2}}.$$

We set g in Equation (6.54) equal to

$$g = \frac{y - xy'}{(x^2 + y^2)[1 + (y')^2]^{1/2}} = \frac{y - xy'}{X^{1/2}\{(x^2 + y^2)[1 + (y')^2]\}^{1/2}}$$

and use Equations (6.53) to obtain

$$g = \frac{\tilde{Y}}{[X(1+\tilde{Y}^2)]^{1/2}}.$$

Since g only depends on X and $\tilde{Y}$, the assumption of rotational invariance is correct. Equation (6.54) simplifies to

$$\frac{d\tilde{Y}}{dX} = -\frac{\tilde{Y}(1+\tilde{Y}^2)}{X},$$

or

$$\frac{dX}{X} = -\frac{d\tilde{Y}}{\tilde{Y}(1+\tilde{Y}^2)},$$

which integrates to

$$X = \frac{a(1+\tilde{Y}^2)^{1/2}}{\tilde{Y}},$$

where a is an integration constant. With the aid of Equations (6.53), this becomes

$$(x^2+y^2)^2\left(\frac{y-xy'}{x+yy'}\right)^2 = a^2\left[1+\left(\frac{y-xy'}{x+yy'}\right)^2\right]. \qquad (6.56)$$

By means of item 6 in Table 5.1, we see that this equation is invariant under the rotation group. Canonical coordinates are introduced via Equations (5.35), with the result

$$\frac{dY}{dX} = \frac{a}{2X(X^2-a^2)^{1/2}},$$

which is separable, as expected. Integration of this equation results in

$$\tan^{-1}(y/x) = b + \frac{1}{2}\sec^{-1}\left(\frac{x^2+y^2}{a}\right),$$

where the angle b is a second integration constant. This relation is the general solution of Equation (6.55). After some algebra (see Problem 6.11), it can be simplified to

$$y^2 - x^2 + bxy + a = 0,$$

where a and b are new constants.

As the final rotation group example, let us apply the theory in Section 6.3 to Equation (6.55). We introduce the notation

$$y_1 = y, \quad y_2 = y'$$

to obtain for Equations (6.33)

$$f_1 = \left(1+y_2^2\right)(y_1 - xy_2) - \left(x^2+y_1^2\right)y_2' = 0 \qquad (6.57)$$
$$f_2 = y_2 - y_1' = 0.$$

By differentiating f_1, Equation (6.35a) becomes

$$-\big[(1 + y_2'^2)\, y_2 + 2xy_2' \big]\xi + (1 + y_2'^2 - 2y_1 y_2')\, \eta_1$$
$$+ (-x + 2y_1 y_2 - 3xy_2'^2)\, \eta_2 - (x^2 + y_1^2)\, \eta_2' = 0. \qquad (6.58)$$

But η_2 is given by Equation (6.37) and η_2' by Equation (4.19), with the result

$$\eta_2' = \eta'' = \eta_{1,xx} + (2\eta_{1,xy_1} - \xi_{xx})y_2 + (\eta_{1,y_1 y_1} - 2\xi_{xy_1})y_2'^2$$
$$- \xi_{y_1 y_1} y_2'^3 + (\eta_{1,y_1} - 2\xi_x - 3\xi_{y_1} y_2)y_2'.$$

With Equation (6.57), y_2' is eliminated from η_2'. The above relation and Equation (6.37) are then substituted into Equation (6.58) to yield

$$-\big[(1 + y_2'^2)y_2 + 2xy_2'\big]\xi + (1 + y_2'^2 - 2y_1 y_2')\, \eta_1$$
$$+ (-x + 2y_1 y_2 - 3xy_2'^2)\big[\eta_{1,x} + (\eta_{1,y_1} - \xi_x)y_2 - \xi_{y_1} y_2'^2\big] - (x^2 + y_1^2)\, \eta_{1,xx}$$
$$- (x^2 + y_1^2)(2\eta_{1,xy_1} - \xi_{xx})\, y_2 - (x^2 + y_1^2)(\eta_{1,y_1 y_1} - 2\xi_{xy_1})\, y_2'^2$$
$$+ (x^2 + y_1^2)\, \xi_{y_1 y_1} y_2'^3 - (\eta_{1,y_1} - 2\xi_x - 3\xi_{y_1} y_2)(1 + y_2'^2)(y_1 - xy_2) = 0. \quad (6.59)$$

The unknown functions in this equation are ξ and η_1, which only depend on x and y_1.

In view of the dependence of ξ and η_1 on x and y_1, the above equation can be decomposed such that the coefficients of

$$1, y_2, y_2^2, y_2^3, y_2'$$

are set equal to zero. (The coefficient of y_2^4 is identically zero, and need not be considered.) This process yields:

$$\eta_1 - x\eta_{1,x} - y_1 \eta_{1,y_1} - (x^2 + y_1^2)\, \eta_{1,xx} + 2y_1 \xi_x = 0 \qquad (6.60a)$$

$$2y_1 \eta_{1,x} - 2(x^2 + y_1^2)\, \eta_{1,xy_1} - \xi - x\xi_x + 3y_1 \xi_{y_1}$$
$$+ (x^2 + y_1^2)\, \xi_{xx} = 0 \qquad (6.60b)$$

$$\eta_1 - 3x\eta_{1,x} + y_1 \eta_{1,y_1} - (x^2 + y_1^2)\, \eta_{1,y_1 y_1} - 2x\xi_{y_1}$$
$$+ 2(x^2 + y_1^2)\, \xi_{xy_1} = 0 \qquad (6.60c)$$

$$2x\eta_{1,y_1} + \xi - x\xi_x - y_1 \xi_{y_1} - (x^2 + y_1^2)\, \xi_{y_1 y_1} = 0 \qquad (6.60d)$$

$$y_1 \eta_1 + x\xi = 0. \qquad (6.60e)$$

These equations constitute an overdetermined system that consists of five, coupled, linear PDEs for ξ and η_1. By means of Equation (6.60e), either ξ or η_1 can be eliminated, thereby yielding a system of four equations for the remaining dependent variable.

A general solution would yield all the groups under which Equation (6.54) is invariant. Of course, we only need one nontrivial, particular solution, which would represent a one-parameter group that would enable us to effect one integration of Equation (6.55). One such solution is provided by the rotation group

$$\xi = -y_1, \quad \eta_1 = x,$$

as can be verified by direct substitution into Equations (6.60). [We know this is a solution, since Equation (6.55) was shown to be invariant under rotation.] In view of their complexity, however, we abandon any thought of trying to systematically solve an overdetermined system such as Equations (6.60).

The next example treats the linear, homogeneous equation of nth order

$$a_n y^{(n)} + a_{n-1} y^{(n-1)} + \cdots + a_0 y = 0, \tag{6.61}$$

where the a_i are functions of x. It is a simple matter to show that this ODE is invariant under the transformation

$$x_1 = x, \quad y_1 = (1 + \alpha)y.$$

As shown by item 4 in Table 5.1, this is an affine transformation whose symbol is

$$Uf = yf_y.$$

To utilize the reduction in order provided by Equation (6.24), we use the new variables

$$X = u = x$$

$$Y = \int \frac{dy}{\eta} = \int \frac{dy}{y} = \ln y. \tag{6.62}$$

With these relations, we readily obtain

$$y' = yY$$
$$y'' = y\left[Y'' + (Y')^2\right]$$
$$y''' = y\left[Y''' + 3Y'Y'' + (Y')^3\right]$$
$$\vdots$$

For instance, if Equation (6.61) is a third-order ODE, it transforms to

$$a_3 \left[z'' + 3zz' + (z')^2\right] + a_2 \left(z' + z^2\right) + a_1 z + a_0 = 0, \tag{6.63}$$

where

$$z(x) = Y'. \tag{6.64}$$

Equation (6.63) is a second-order, inhomogeneous, nonlinear equation. We thus see that any homogeneous, linear ODE (when $n \geq 2$) can be transformed by Equations (6.62) to a nonlinear, inhomogeneous ODE whose order is reduced by one. Equation (6.64) is solved after the new ODE is solved. This results in the quadrature

$$Y = \ln b + \int z(x)dx,$$

or

$$y = b \exp\left(\int z dx\right),$$

where b is an integration constant.

In the next example, we find the ODE invariant under an enlarged item IIa2 group when

$$p = X.$$

From Table 5.4, we have

$$X = \frac{b}{a}x - y = \frac{1}{a}(bx - ay)$$
$$Uf = af_x + bf_y$$

so that the enlarged group symbol is

$$\widehat{U}f = \frac{1}{a}(bx - ay)(af_x + bf_y). \tag{6.65}$$

With

$$q = -\frac{1}{p}\frac{dp}{dX} = -\frac{1}{X},$$

Equation (6.47) becomes

$$\frac{d^2Y}{dX^2} - \frac{2}{X}\frac{dY}{dX} + 2\frac{Y}{X^2} = g\left(X, \frac{dY}{dX} - \frac{Y}{X}\right). \tag{6.66a}$$

In this relation, X, Y, Y', and Y'' are given by item IIa2 in Tables 5.4 and 6.1, with the result

$$\frac{a^3y''}{(b - ay')^3} - \frac{2a^2}{(bx - ay)(b - ay')} + \frac{2a^2x}{(bx - ay)^2} = g\left(bx - ay, \frac{a^2(xy' - y)}{(bx - ay)(b - ay')}\right).$$

The second argument of g simplifies to

$$\frac{xy' - y}{b - ay'}.$$

The second and third terms on the left side combine to

$$\frac{2a^3(xy' - y)}{(bx - ay)^2(b - ay')},$$

which can be absorbed into g. After further simplification, Equation (6.66a) finally reduces to

$$y'' = (xy' - y)^3 g\left(bx - ay, \frac{xy' - y}{b - ay'}\right), \tag{6.66b}$$

which is invariant under the group represented by Equation (6.65). This relation should be compared with item IIa2 in Table 6.1 for the original group. Since $p = X$, the introduction of Equations (5.49)

$$X = bx - ay$$
$$\widehat{Y} = \frac{1}{bx - ay}\frac{xy' - y}{b - ay'}$$

reduces Equation (6.66b) to first order, i.e.,

$$\frac{d\widehat{Y}}{dX} = g(X, \widehat{Y}).$$

The specific form for this equation is part of Problem 6.10.

In quantum mechanics, the eigenfunctions and eigenvalues stem from solutions of the *Sturm-Liouville equation*

$$y'' + \left[\lambda^2 - f(x)\right] y = 0 \qquad (6.67)$$

where λ is a constant, f is a known function, and $y \to 0$ as $x \to \infty$. This equation arises when a separation of variable solution is obtained for the Schrodinger wave equation. Solutions of the above equation are well known. For example, for a simple harmonic oscillator, with nondimensional variables,

$$f = x^2$$

and the solution is

$$y_\nu = e^{-x^2/2} H_\nu(x), \quad \nu = 0, 1, \ldots,$$

where the integer ν is the vibrational quantum number and the *Hermite polynomials* are given by

$$H_\nu = (-1)^\nu e^{x^2} \frac{d^\nu}{dx^\nu} e^{-x^2}.$$

In this example, Equation (6.67) is reduced to first-order by showing that it is invariant under item Ias1. This is done by setting

$$A = y$$

$$g\left(x, \frac{y'}{y}\right) = f(x) - \lambda^2 - \left(\frac{y'}{y}\right)^2$$

in the Table 6.1 ODE column. From Tables 5.4 and 6.1, we use

$$X = x, \quad Y = \ln y$$

$$Y' = \widehat{Y} = \frac{y'}{y}$$

$$Y'' = \widehat{Y}' = \frac{yy'' - (y')^2}{y^2}$$

to obtain a Riccati equation

$$\widehat{Y}' + \widehat{Y}^2 = f(X) - \lambda^2$$

for the first-order version of Equation (6.67). This is not surprising, since every second-order, linear ODE can be transformed into a Riccati equation (Rainville, 1943).

The next two examples are based on Bluman and Cole (1974). In the first, the ODE for the shape of an ogive

$$y = F(x)$$

is examined that results in minimum drag for a vehicle in hypersonic flow. The analysis leading to the ODE assumes a Newtonian impact pressure relation and a laminar boundary layer. From variational calculus, an Euler-Lagrange equation for the shape yielding the minimum drag

$$3yy'y'' + (y')^3 - kx^{-1/2} = 0 \qquad (6.68)$$

is obtained, where k is a constant. This equation can be written as

$$y'' = -\frac{(y')^2}{3y} + \frac{k}{3yy'x^{1/2}},$$

or

$$x^2 y'' = y\left[-\frac{1}{3}\left(\frac{xy'}{y}\right)^2 + \frac{k\,x^{5/2}}{3\,y^3}\left(\frac{y}{xy'}\right)\right]. \qquad (6.69a)$$

This relation falls under item IIIa3 in Table 6.1, with

$$a = 3, \quad b = \frac{5}{2}, \quad g = \frac{1}{3}\left[k\left(\frac{x^{5/2}}{y^3}\right)\left(\frac{y}{xy'}\right) - \left(\frac{xy'}{y}\right)^2\right].$$

The reduction in order is achieved by utilizing from Table 5.4

$$X = \frac{x^b}{y^a} = \frac{x^{5/2}}{y^3}$$

$$Y' = \frac{dY}{dX} = \frac{y^4}{x^{5/2}\left(\frac{5}{2}y - 3xy'\right)} = \left(\frac{y^3}{x^{5/2}}\right)\frac{1}{\frac{5}{2} - 3\left(\frac{xy'}{y}\right)}.$$

In accord with the second argument of g in Table 6.1, it is more convenient to use

$$\widetilde{Y} = \frac{xy'}{y} \qquad (6.70)$$

in place of Y'. Equation (6.69a) becomes

$$y'' = \frac{y}{3x^2}\left(\frac{kx}{\widetilde{Y}} - \widetilde{Y}^2\right). \qquad (6.69b)$$

We now determine $d\widetilde{Y}/dX$ as follows:

$$\frac{dX}{dx} = \frac{5}{2}\frac{x^{3/2}}{y^3} - \frac{3x^{5/2}}{y^4}y' = \frac{X}{x}\left(\frac{5}{2} - 3\widetilde{Y}\right)$$

$$\frac{d\widetilde{Y}}{dX} = \frac{dx}{dX}\frac{yy' + xyy'' - x(y')^2}{y^2} = \frac{x[xyy'' - x(y')^2 + yy']}{y^2 X\left(\frac{5}{2} - 3\widetilde{Y}\right)}.$$

The y' and y'' variables are replaced using the above relations, with the final result

$$\frac{d\widetilde{Y}}{dX} = \frac{2kX - 8\widetilde{Y}^2 + 6\widetilde{Y}}{3X\widetilde{Y}(5 - 6\widetilde{Y})}. \tag{6.69c}$$

This is the first-order replacement for Equation (6.68). It does not appear to be invariant under any group, although it is readily solved when $k = 0$. However, this parameter value is physically unrealistic, since it corresponds to a vehicle without drag. Even after this equation is solved, a second integration is required to go from $X, \widetilde{Y}$ to x, y variables.

In atomic physics, the Thomas-Fermi theory is used to obtain an effective potential in a solid due to the orbiting electrons in an atom. This potential is then used to derive a thermal equation of state for the atom, which is located in a solid. In an approximate version of the theory (Bluman and Cole, 1974), the equation

$$y'' = x^{-1/2}y^{3/2} \tag{6.71}$$

in encountered. (An exact version of the theory can be found in Latter, 1955.) By writing this equation as

$$x^2 y'' = y\left(x^{3/2}y^{1/2}\right),$$

we see that it too is covered by item IIIa3, with

$$a = -\frac{1}{2}, \quad b = \frac{3}{2}, \quad X = x^{3/2}y^{1/2}, \quad g = X.$$

From Table 6.1, we have

$$\widetilde{Y} = \frac{x}{y}y'$$

and, consequently, Equation (6.71) becomes

$$\frac{dX}{dx} = \frac{3}{2}x^{1/2}y^{1/2} + \frac{x^{3/2}y'}{2y^{1/2}} = \frac{X}{2x}\left(3 + \widetilde{Y}\right)$$

$$\frac{d\widetilde{Y}}{dX} = \frac{2(X - \widetilde{Y}^2 + \widetilde{Y})}{X(3 + \widetilde{Y})}.$$

This equation does not appear to be invariant under a known group.

6.6 Problems

6.1. Find the second-order ODE invariant under the third one-parameter group in Problem 2.2.

6.2. Repeat Problem 6.1 for the second one-parameter group in Problem 2.2.

6.3. Find the solution to

$$x^5 y'' = xy^2 y' - y^3$$

subject to the boundary condition

$$y' = 0 \quad \text{when } y = 0$$

in the form

$$y = f(x, b),$$

where b is a constant of integration.

6.4. Solve

$$y'' = y^2 + y^3 + 2\left(-\frac{1}{x} + xy + xy^2\right) y' + (1 + y)x^2 (y')^2$$

in a form that involves two integration constants. (Hint: Use the results of Problem 6.2.)

6.5. Find the general form of the second-order ODE invariant under the group

$$x_1 = x \cosh \alpha + y \sinh \alpha$$
$$y_1 = x \sinh \alpha + y \cosh \alpha.$$

6.6. In one spatial dimension, the Lorentz-Einstein transformation of special relativity is given by

$$x_1 = \phi(x, t, V) = \frac{x - Vt}{[1 - (V/c)^2]^{1/2}}$$

$$t_1 = \psi(x, t, V) = \frac{t - (V/c^2)x}{[1 - (V/c)^2]^{1/2}},$$

where c is the constant speed of light. This transformation is a continuous Lie group where the speed, V, is the group parameter. Determine the most general form of the first- and second-order ODEs invariant under this group. Aside from constants, your answers should only involve t, x, dx/dt, and d^2x/dt^2.

6.7. What are the finite, or global, equations of the group whose symbol is

$$Uf = \left(x - \frac{1}{2x}\right) f_x + \left(y - \frac{1}{2y}\right) f_y.$$

Determine the general form of a first-order ODE invariant under the group, the canonical variables, and the separable form, Equation (5.33). Finally, determine the general form of the second-order ODE invariant under the group. Simplify all results and use Table 6.1 to check the second-order ODE result.

6.8. Use the third method, given by Equation (6.23), to derive the second-order ODE invariant under the rotation group. Simplify your result and show that it is consistent with Equation (6.50a).

6.9. Solve the following ODEs:

$$xyy'' + x(y')^2 - yy' = 0 \tag{1}$$
$$yy'' + (y')^2 = 1 \tag{2}$$
$$y'' = (y')^2 + 1 \tag{3}$$
$$x^2yy'' - (xy' - y)^2 = 0 \tag{4}$$
$$(x^2 + y^2)y'' + 2(y - xy')[1 + (y')^2] = 0 \tag{5}$$
$$x^2y'' + x^2(y')^2 - 2xy' + 2 = 0. \tag{6}$$

Without exception, all answers should have the form

$$y = y(x; a, b),$$

where a and b are two integration constants and the algebraic function on the right side is free of quadratures.

6.10. Suppose g is given by

$$g = (bx - ay)^2 \left(\frac{xy' - y}{b - ay'} \right)^3.$$

Integrate Equation (6.66b) once with this g to obtain a first-order ODE involving x, y, and y'.

6.11. Show that the solution of Equation (6.55)

$$\tan^{-1}(y/x) = b + \frac{1}{2} \sec^{-1} \left(\frac{x^2 + y^2}{a} \right)$$

can be reduced to

$$y^2 - x^2 + bxy + a = 0$$

where a and b are new constants, and verify the result.

6.12. Find the general solution to

$$yy'' - x^2y^2(y')^2 - xy^3y' - yy' - x = 0.$$

6.13. Use Table 6.1 to derive the ODE

$$y'' = [1 + (y')^2]^{3/2} g \left(x^2 - y^2, \ (x + y)^2 \frac{1 - y'}{1 + y'} \right)$$

that is invariant under the group

$$Uf = yf_x + xf_y.$$

6.14. Derive item IIIa3 in Table 6.1.

6.15. Use item IIIcs4 in Table 6.1 to show that the second-order ODE invariant under

$$Uf = x^{n+1}f_x + nx^n y f_y$$

can be written as

$$x^{n+2}y'' + (1-n)x^{n+1}y' = g(y/x^n, xy' - ny).$$

What ODE does the substitution principle yield?

6.16. Prove that

$$\frac{d^2 u^{(1)}}{du^2}$$

is a third differential invariant.

6.17. Find the general solution to

$$y'' - (x-y)(y')^3 = 0.$$

6.18. Use the theory in Section 6.2 to determine a fourth-order ODE for ξ whose solution is required by the invariance of

$$y'' = g_o(x)y^2.$$

6.19. Utilize Equations (6.28) to determine the PDEs that yield the groups for which

$$y'' = g_1(x, y)y'$$

is invariant. Find the functional form for $\xi(x, y)$ and determine two PDEs for η that are free of g_1 and its derivatives.

6.20. For the third-order ODE

$$f(x, y, y', y'', y''') = 0,$$

determine the equations that are comparable to Equations (6.35) and (6.36).

6.21. Derive the equivalent of Equation (6.69c) using Y' instead of $\tilde{Y}$.

6.22. Derive the two ODEs for the $a_i, i = 1, \ldots, 4$, mentioned in the next to last paragraph in Section 6.2.

6.23. Use the decomposition method of Section 6.2 to determine the independent groups under which

$$y'' = e^{xy}$$

is invariant.

Chapter 7

Second-Order ODEs Invariant Under a Two-Parameter Group

Sections 7.1 and 7.2 provide a relatively complete outline of the theory, in part, because of its greater complexity as compared to the theory in the previous two chapters. These sections cover the definition and classification of two-parameter groups, invariance, and canonical coordinates. We shall find that every second-order ODE invariant under a two-parameter group can be fully solved in terms of quadratures. This is achieved by determining the classification, or type, for the group and introducing appropriate canonical coordinates that result in a separable form for the ODE which is twice integrable.

The one-parameter theory of Chapter 6 does not generally provide a full solution for a second-order ODE. In this regard, the methods of this chapter are clearly preferable. Despite this, the two-parameter theory has previously been limited in its practical application. This limitation is due to the lack of general symbols, such as those in Table 5.2, the lack of comprehensive tables, such as Tables 5.4 and 6.1, and the complexity of the theory. These deficiencies are partly rectified in Section 7.3, which also discusses the substitution principle and the enlargement procedure. Overall, this section increases the scope of the theory, by increasing the number of ODEs subject to solution, and simplifies its method of application.

The chapter concludes with a variety of examples in Section 7.4.

7.1 Classification of Two-Parameter Groups

Commutator

Our first task is to define a two-parameter group. When two symbols are under consideration, their *commutator* first needs to be introduced. This operator of U_1 and U_2, written as $(U_1 U_2)f$, is defined by

$$(U_1 U_2)f = U_1(U_2 f) - U_2(U_1 f). \tag{7.1}$$

By writing this relation for $(U_2 U_1)f$, we observe that if U_1 and U_2 commute, then the commutators, $(U_1 U_2)$ and $(U_2 U_1)$, are both zero. Thus, a commutator represents two operators that don't commute. Commutators appear in many areas of mathematics and physics; they are especially important when dealing with angular momentum in quantum mechanics.

With the U_i written as

$$U_i f = \xi_i f_x + \eta_i f_y, \quad i = 1, 2, \tag{7.2}$$

we have

$$
\begin{aligned}
U_1(U_2 f) &= \xi_1 \frac{\partial}{\partial x}(\xi_2 f_x + \eta_2 f_y) + \eta_1 \frac{\partial}{\partial y}(\xi_2 f_x + \eta_2 f_y) \\
&= \xi_1(\xi_{2x} f_x + \eta_{2x} f_y) + \xi_1(\xi_2 f_{xx} + \eta_2 f_{xy}) \\
&\quad + \eta_1(\xi_{2y} f_x + \eta_{2y} f_y) + \eta_1(\xi_2 f_{xy} + \eta_2 f_{yy}) \\
&= f_x U_1 \xi_2 + f_y U_1 \eta_2 + \xi_2 U_1 f_x + \eta_2 U_1 f_y. \tag{7.3a}
\end{aligned}
$$

Similarly, we obtain

$$U_2(U_1 f) = f_x U_2 \xi_1 + f_y U_2 \eta_1 + \xi_1 U_2 f_x + \eta_1 U_2 f_y \tag{7.3b}$$

by interchanging the 1 and 2 subscripts. We thereby obtain

$$(U_1 U_2)f = (U_1 \xi_2 - U_2 \xi_1)f_x + (U_1 \eta_2 - U_2 \eta_1)f_y, \tag{7.4}$$

since the terms

$$\xi_2 U_1 f_x - \xi_1 U_2 f_x + \eta_2 U_1 f_y - \eta_1 U_2 f_y = 0$$

sum to zero. Thus, all terms containing second derivatives of f cancel. Hereafter, Equation (7.4) is used whenever the commutator is to be evaluated. For this, we merely determine the symbols $U_i \xi_j$ and $U_i \eta_j$.

By interchanging digital subscripts in Equation (7.4), we observe

$$(U_1 U_2)f = -(U_2 U_1)f.$$

This antisymmetric property, however, plays no role in the subsequent analysis. If the U_i are *linearly dependent*, then

$$U_2 f = e U_1 f,$$

where e is a nonzero constant. If no such relation exists, the two groups are *linearly independent*. When the groups are linearly dependent, the commutator is

$$(U_1 U_2)f = U_1(e U_1 f) - e U_1(U_1 f) = 0,$$

and $(U_1U_1)f$ is zero. The commutator of linearly independent groups may also be zero. This point is verified in the next section.

Definition of a Multi-parameter Group

A *two-parameter group* consists of two, one-parameter groups whose symbols are given by Equation (7.2). If the groups *jointly* satisfy the Section 2.1 group definitions, they would then constitute a single group. A necessary and sufficient condition for this property is that the commutator has the form

$$(U_1U_2)f = e_1U_1f + e_2U_2f, \tag{7.5}$$

where the e_i are constants and Equation (7.4) holds for the commutator. The two U_if symbols must be linearly independent; they represent a *basis* for a two-parameter group.

Not every combination of one-parameter groups satisfies the above equation. For example, the symbols

$$U_1f = f_y, \quad U_2f = (\ln y)f_y$$

yield

$$(U_1U_2)f = \frac{1}{y}f_y,$$

which cannot equal $(e_1 + e_2 \ln y)f_y$. These two symbols do not constitute a two-parameter group.

Groups with three or more parameters are possible. In this case, the linearly independent one-parameter groups, $U_if, i = 1, \ldots, r$, where $r \geq 2$, constitute an r-parameter group if every two distinct U_if satisfy the commutator relation

$$(U_iU_j)f = \sum_{k=1}^{r} e_{ijk}U_kf, \quad i, j = 1, \ldots, r,$$

where the e_{ijk} are constants.

First Classification

Without loss of generality, one can show that every two-parameter group can be made to satisfy either

$$(U_1U_2)f = 0 \tag{7.6a}$$

or

$$(U_1U_2)f = U_1f. \tag{7.6b}$$

Thus, the e_i in Equation (7.5) can be chosen as

$$e_1 = e_2 = 0$$

for Equation (7.6a), or as

$$\epsilon_1 = 1, \quad \epsilon_2 = 0$$

for Equation (7.6b). (In vector analysis, this is equivalent to choosing a basis that spans the space. One difference, however, is that the null vector can be chosen in lieu of the $e_1 = 0$, $e_2 = 1$ vector.) If one of Equations (7.6) is not satisfied, a linear transformation to a new symbol pair is used. The transformation is written as

$$V_1 f = \epsilon_{11} U_1 f + \epsilon_{12} U_2 f$$
$$V_2 f = \epsilon_{21} U_1 f + \epsilon_{22} U_2 f, \tag{7.7}$$

where the ϵ_{ij} are constants and the $V_i f$ symbols are linearly independent.

With the aid of Equations (7.4), one can show that the commutators of the V and U symbols are related by (see Problem 7.8)

$$(V_1 V_2) f = \Delta (U_1 U_2) f,$$

where

$$\Delta = e_{11} e_{22} - e_{12} e_{21}.$$

The transformed counterpart to Equation (7.5) is

$$(V_1 V_2) f = \bar{e}_1 V_1 f + \bar{e}_2 V_2 f,$$

where the $\bar{e}_i$ satisfy either

$$\bar{e}_1 = \bar{e}_2 = 0$$

or

$$\bar{e}_1 = 1, \quad \bar{e}_2 = 0.$$

With the foregoing relations, we obtain

$$(\bar{e}_1 V_1 f + \bar{e}_2 V_2 f) = \Delta (e_1 U_1 f + e_2 U_2 f),$$

which becomes, with $\bar{e}_2 = 0$,

$$\bar{e}_1 V_1 f = \bar{e}_1 (\epsilon_{11} U_1 f + \epsilon_{12} U_2 f) = \Delta (e_1 U_1 f + e_2 U_2 f)$$

or

$$(\bar{e}_1 \epsilon_{11} - e_1 \Delta) U_1 f + (\bar{e}_1 \epsilon_{12} - e_2 \Delta) U_2 f = 0.$$

The coefficients of the $U_i f$ are set equal to zero, with the result

$$\bar{e}_1 \epsilon_{11} = e_1 \Delta, \quad \bar{e}_1 \epsilon_{12} = e_2 \Delta \tag{7.8}$$

for the e_{ij} in Equations (7.7). In addition, a value for Δ is chosen such that $\bar{e}_1$ is zero or unity.

As an example, suppose we have

$$U_1 f = x f_x, \quad U_2 f = x f_y,$$

which yields for the commutator

$$(U_1U_2)f = U_2f.$$

This result is not in accord with Equations (7.6). To obtain transformed symbols that are in accord, set

$$e_1 = 0, \quad e_2 = 1, \quad \bar{e}_1 = 1, \quad \bar{e}_2 = 0$$

in Equations (7.8) along with $\Delta = 1$. We, thereby, obtain

$$e_{11} = 0, \quad e_{21} = -1, \quad e_{12} = 1, \quad e_{22} = 0,$$

where the value for e_{22} is arbitrary. We thus have

$$V_1f = U_2f = xf_y$$
$$V_2f = -U_1f = -xf_x$$

and the commutator

$$(V_1V_2)f = xf_y = V_1f$$

has $\bar{e} = 1$, $\bar{e}_2 = 0$ as required.

Second Classification and Fundamental Forms

A second classification of two-parameter groups depends on the function

$$\phi = U_2f - \rho(x,y)U_1f, \tag{7.9}$$

where ρ is not a constant. There are two cases. In the first, ρ is a function of x and y such that ϕ is identically zero. In the second, $\phi \neq 0$ for all possible choices of $\rho(x,y)$.

The two modes of classification are independent of each other. Consequently, every two-parameter group falls into one of four possible *types*. These are summarized in Table 7.1, where this classification is not to be confused with those in earlier chapters. Each type is invariant under a linear transformation of the symbols. Thus, the symbols of type I remain type I under such a transformation.

Table 7.1 shows the e_i along with the *fundamental*, or *standard*, *forms for the commutator* and the function ϕ. If the U_if are such that the e_i are not in accord with those in the table, then a linear transformation of the U_if is used to bring the V_if into conformity.

7.2   Invariance and Canonical Coordinates

Invariant ODE

Recall from Section 6.1 that a second-order ODE is invariant under a one-parameter group if Equation (6.2) holds. In this circumstance, the ODE can be

written as the first of Equations (6.7). These statements now hold individually
for each of the $U_i f$ symbols, with the result

$$G_i \left(u_i, u_i^{(1)}, u_i^{(2)} \right) = 0, \quad i = 1, 2. \tag{7.10}$$

For a second-order ODE to be invariant under a two-parameter group, the
two $u_i^{(2)}$ must be equivalent. In accord with Equation (6.8), we have that the
$u_i^{(2)}$ are functionally related by

$$u_1^{(2)} = F_1 \left(u_2, u_2^{(1)}, u_2^{(2)} \right) \tag{7.11a}$$

or

$$u_2^{(2)} = F_2 \left(u_1, u_1^{(1)}, u_1^{(2)} \right). \tag{7.11b}$$

In addition, we also must have u_1 equivalent to u_2 or, alternatively, $u_1^{(1)}$ and
$u_2^{(1)}$ equivalent. However, all three $U_1 f$ invariants *cannot* be equivalent with all
three from $U_2 f$ if the two symbols are to be linearly independent. However, the
equivalence of the $u_i^{(2)}$ invariants is essential, since y'' only appears in the second
differential invariants. When these conditions are satisfied, the two-parameter
group admits a second-order ODE. This ODE can be written as

$$G \left(\hat{u}, u^{(2)} \right) = 0, \tag{7.12}$$

where $u^{(2)}$ is equivalent to both of the $u_i^{(2)}$ and $\hat{u}$ is equivalent to u_i or to $u_i^{(1)}$.

Canonical Form of the ODE

Not only does every two-parameter group correspond to a second-order ODE,
but the ODE can be fully solved in terms of quadratures. The key to this step
is to transform the ODE into a separable, or *canonical form*. Table 7.2 shows
these forms in its last column (Axford, 1971). As is evident from the symbol
column, these forms correspond to the four fundamental commutator forms of
Table 7.1. For instance, the $U_i f$ of type I in Table 7.2 results in a zero value
for the commutator and $\phi \neq 0$. Also shown are the invariants and the first and
second differential invariants for each of the eight symbols.

As a review, let us first verify that the type III symbols in Table 7.2 are in
accord with Table 7.1. For these symbols

$$U_1 f = f_y, \quad U_2 f = x f_x + y f_y \tag{7.13}$$

we have

$$\xi_1 = 0, \quad \eta_1 = 1, \quad \xi_2 = x, \quad \eta_2 = y,$$

and consequently,

$$U_1\xi_2 = \xi_1 \frac{\partial x}{\partial x} + \eta_1 \frac{\partial x}{\partial y} = 0$$

$$U_2\xi_2 = 0$$

$$U_1\eta_2 = 1$$

$$U_2\eta_1 = 0.$$

The commutator, Equation (7.4), thus becomes

$$(U_1U_2)f = f_y = U_1f,$$

and ϕ is

$$\phi = xf_x + (y - \rho)f_y.$$

Since no single function, $\rho(x, y)$, enables ϕ to be identically zero, the type III classification for Equations (7.13) is confirmed.

Next, we verify the type III invariants and the final column in Table 7.2 for Equations (7.13). By means of Appendix C, we obtain

$$\eta_1' = \eta_{1x} + (\eta_{1y} - \xi_{1x})y' - \xi_{1y}(y')^2 = 0$$

$$\eta_2' = 0$$

$$\eta_1'' = 0$$

$$\eta_2'' = -y''.$$

Hence, Equations (6.4) become

$$\frac{dx}{0} = \frac{dy}{1} = \frac{dy'}{0} = \frac{dy''}{0} \tag{7.14a}$$

$$\frac{dx}{x} = \frac{dy}{y} = \frac{dy'}{0} = -\frac{dy''}{y''} \tag{7.14b}$$

for the two, twice-extended symbols. We thus obtain

$$u_1(x, y) = x$$

$$u_1^{(1)}(x, y, y') = y' \tag{7.15a}$$

$$u_1^{(2)}(x, y, y', y'') = y'',$$

which may be verified by noting that each of the above invariants equals a constant of integration. From Equations (7.14b), we have

$$u_2(x, y) = \frac{y}{x}$$

$$u_2^{(1)}(x, y, y') = y' \tag{7.15b}$$

$$u_2^{(2)}(x, y, y', y'') = xy''.$$

To verify $u_2^{(2)}$, set

$$xy'' = \text{constant}$$

and obtain by logarithmic differentiation

$$\frac{dx}{x} + \frac{dy''}{y''} = 0,$$

which is in accord with Equations (7.14b). Thus, Equations (6.7) are

$$G_1(x, y', y'') = 0$$
$$G_2\left(\frac{y}{x}, y', xy''\right) = 0 \qquad\qquad (7.16)$$

for the two, one-parameter groups of type III. Observe that G_1 also can be written as

$$G_1(x, y', xy'') = 0$$

and by comparison with G_2, we obtain

$$xy'' = g(y'), \qquad\qquad (7.17)$$

which is the result given in the last column of Table 7.2.

Integration of the Canonical ODEs

Each of the four canonical equations listed in the last column of Table 7.2 is separable. Thus, each equation can be integrated once in terms of quadratures. In principle, each equation actually can be twice integrated in terms of quadratures.

As an example, this assertion is verified for Equation (7.17). Thus, one integration yields

$$x\frac{dy'}{dx} = g(y')$$
$$\frac{dy'}{g(y')} = \frac{dx}{x}$$
$$\int \frac{dy'}{g(y')} = \ln x + \ln a = \ln(ax),$$

where a is a constant of integration. Define a function $e(y')$ by means of

$$\ln e(y') = \int \frac{dy'}{g(y')}$$

so that

$$e(y') = ax.$$

By inversion, this relation becomes

$$y' = h(ax),$$

which integrates to

$$y = \frac{1}{a} \int h(ax)d(ax) + b.$$

This is the general solution to Equation (7.17).

For instance, if g equals

$$g = y' \ln y',$$

then

$$e = \ln y'$$

and the solution is

$$y = \frac{1}{a} e^{ax} + b.$$

In this case, an explicit form for the solution is found. However, if $g = (\ln y')^{-1}$, then $e(y)$ is given by

$$\ln e = y'(\ln y' - 1).$$

This relation cannot be inverted, and $h(ax)$ and an explicit form for the solution cannot be found. Of course, a numerical solution can be obtained.

Canonical Coordinates

By way of summary, suppose we have two, one-parameter groups whose commutator satisfies Equation (7.5). We thus have a two-parameter group, which, by means of a linear symbol transformation, if necessary, coincides with one of the four types in Table 7.1. The primary reason for determining the type will become apparent shortly, when discussing canonical coordinates. We next determine Equation (6.7), (6.11), or (6.23) for each of the (transformed) symbols. After this, a second-order ODE, Equation (7.12), is formed. [Later sections provide alternative procedures for determining Equation (7.12).] Let us now presume that this equation has been found, but that it is not separable. We can transform it into a separable form by introducing canonical coordinates. After the transformation, the resulting ODE will have the functional form given in Table 7.2 that holds for the type of the original equation. Thus, the coordinate transformation preserves the type of the original ODE.

Two somewhat different transformations are required (Cohen, 1911). The first is a direct generalization of that in Section 5.2, but holds only for types I and III, where $\phi \neq 0$. In this case, canonical coordinates X and Y are introduced by means of Equations (5.20). For types I and III, these coordinates are determined by four PDEs

$$U_i X = \bar{\xi}_i, \quad U_i Y = \bar{\eta}_i, \quad i = 1, 2, \tag{7.18}$$

where the $\bar{\xi}_i$ and $\bar{\eta}_i$ are provided by the symbol column in Table 7.2. For convenience, the transformation equations for all types are summarized in Table 7.3. Observe that the types II and IV transformations differ from Equations (7.18).

The use of canonical coordinates is crucial for obtaining a solution in terms of quadratures. As evident in Table 7.3, the transformation equations are type dependent. Thus, knowing the type under which the group falls in is usually a first essential step.

Equations (7.18) contain two algebraic equations for X_x and X_y. They can be solved, with the result

$$X_x = \left(\eta_2 \bar{\xi}_1 - \eta_1 \bar{\xi}_2 \right) / \Delta, \quad X_y = - \left(\xi_2 \bar{\xi}_1 - \xi_1 \bar{\xi}_2 \right) / \Delta, \qquad (7.19a)$$

where

$$\Delta = \xi_1 \eta_2 - \xi_2 \eta_1. \qquad (7.19b)$$

(Don't confuse this Δ with the earlier one.) Similar equations hold for Y_x and Y_y. For types I and III, the determinant Δ is not zero. The reason for a different transformation for types II and IV is that Δ is zero, since $\xi_1 = \xi_2 = 0$, and Equations (7.19a) cannot be used.

Equations (5.20) and (7.19a) imply

$$dX = X_x dx + X_y dy$$
$$= \frac{1}{\Delta} \left(\eta_2 \bar{\xi}_1 - \eta_1 \bar{\xi}_2 \right) dx - \frac{1}{\Delta} \left(\xi_2 \bar{\xi}_1 - \xi_1 \bar{\xi}_2 \right) dy,$$

which yields, upon integration,

$$X(x,y) = \int \left(\eta_2 \bar{\xi}_1 - \eta_1 \bar{\xi}_2 \right) \frac{dx}{\Delta} - \int \left(\xi_2 \bar{\xi}_1 - \xi_1 \bar{\xi}_2 \right) \frac{dy}{\Delta}. \qquad (7.20)$$
$$y = \text{constant} \qquad\qquad x = \text{constant}$$

There is a similar equation for $Y(x,y)$. We shall not write it, since it and the above equation are usually not utilized. This is because Equations (7.18) are generally solved by inspection. Further simplifying our task is the fact that only a particular solution of these equations is required. Indeed, the simplest particular solution is desired, since it results in the simplest coordinate transformation. As with a first-order ODE, the equations for the coordinate transformation are not unique.

Separable Form for the ODE

Suppose a one-to-one transformation, Equations (5.20), has been found for any of the four types. We now develop general relations for transforming Equation (7.12) into its separable form. For this, the inverse of Equations (5.20) is utilized. If these are explicitly available, then the following lengthy procedure is unnecessary. However, Equations (5.20) are not always algebraically invertible, in which case we can proceed as follows. Differentiate Equations (5.20), with the result

$$dX = X_x dx + X_y dy = dx(X_x + X_y y') \qquad (7.21a)$$
$$dY = Y_x dx + Y_y dy = dx(Y_x + Y_y y'). \qquad (7.21b)$$

We thereby obtain

$$Y' = \frac{dY}{dX} = \frac{Y_x + Y_y y'}{X_x + X_y y'},$$ (7.22)

or by solving for y',

$$y' = \frac{Y_x - X_x Y'}{-Y_y + X_y Y'}.$$ (7.23)

From Equation (7.22) we see that Y' is a function of x, y, and y'. Hence, we have

$$dY' = Y'_x dx + Y'_y dy + Y'_{y'} dy',$$

where

$$Y'_x = \frac{\partial}{\partial x}\left(\frac{dY}{dX}\right), \dots$$

Next, divide through by dx, with the result

$$\frac{dY'}{dx} = Y'_x + Y'_y y' + Y'_{y'} y''$$

$$Y'' = \frac{d^2 Y}{dX^2} = \frac{dY'}{dX} = \frac{dx}{dX}(Y'_x + Y'_y y' + Y'_{y'} y'').$$

With the use of Equations (7.21a) and (7.22), we have

$$Y'' = \frac{1}{X_x + X_y y'}\left(\frac{\partial}{\partial x} + y'\frac{\partial}{\partial y} + y''\frac{\partial}{\partial y'}\right)\left(\frac{Y_x + Y_y y'}{X_x + X_y y'}\right).$$ (7.24)

Introduce the notation

$$D = \frac{\partial}{\partial x} + y'\frac{\partial}{\partial y} + y''\frac{\partial}{\partial y'}$$

$$u = Y_x + Y_y y'$$

$$v = X_x + X_y y'$$

so that

$$Y'' = \frac{v\,Du - u\,Dv}{v^3}.$$

Expanded out, this yields

$$Y'' = \frac{1}{v^3}\left[\begin{vmatrix} X_x & X_{xx} \\ Y_x & Y_{xx} \end{vmatrix} + \left(2\begin{vmatrix} X_x & X_{xy} \\ Y_x & Y_{xy} \end{vmatrix} + \begin{vmatrix} X_y & X_{yy} \\ Y_y & Y_{yy} \end{vmatrix}\right)y'\right.$$

$$+ \left(2\begin{vmatrix} X_y & X_{xy} \\ Y_y & Y_{xy} \end{vmatrix} + \begin{vmatrix} X_x & X_{yy} \\ Y_x & Y_{yy} \end{vmatrix}\right)(y')^2 + \begin{vmatrix} X_y & X_{yy} \\ Y_y & Y_{yy} \end{vmatrix}(y')^3$$

$$+ \left.\begin{vmatrix} X_x & X_y \\ Y_x & Y_y \end{vmatrix}y''\right].$$ (7.25)

This equation is the desired result. It requires that x and y be replaced by X and Y, y' is replaced by Equation (7.23), and y'' is replaced using Equation (7.12). Observe that the coefficient of y'' is the Jacobian of the transformation, which is not zero. The above equation results in a separable form for Equation (7.12) of the proper type.

Discussion

Some of the discussion in the literature (Dickson, 1923; Franklin, 1928) concerning the applicability of Equations (7.18) to the types II and IV equations is erroneous. As an example, consider the linearly independent symbols (Franklin, 1928)

$$U_1 f = f_x + j(x, y) f_y, \quad U_2 f = x U_1 f. \tag{7.26}$$

These satisfy type IV with $\phi = 0$ and $\rho = x$. However, Equations (7.18) and the symbol column in Table 7.2 yield the incompatible equations

$$Y_x + j Y_y = 1$$
$$Y_x + j Y_y = \frac{y}{x}$$

for Y.

The reason for this difficulty is that $\Delta = 0$. As can be seen from Equation (7.19b), this problem is not associated with the choice of a separable form, which depends only on the choice of the values for $\bar{\xi}_i$ and $\bar{\eta}_i$. Rather, it depends on the two-parameter group, where the two linearly independent symbols may still yield $\Delta = 0$. In Franklin (1928), the transformation $X = \rho(x, y)$, $Y = y$ is utilized in an attempt to find a quadrature solution. For Equations (7.26), however, the transformation

$$X = \rho = \frac{\bar{\xi}_2}{\bar{\xi}_1} = x, \quad Y = y$$

is the identity transformation, and the analysis is incorrect.

Let us continue with our discussion of Equations (7.26) with $j = x$. From item IIIcs2 in Tables 5.4 and 6.1, we have for the $U_1 f$ symbol

$$a = 1, \quad m = 0, \quad b = 1, \quad n = 1, \quad r = 0,$$

and

$$xyy'' - xy = xyg_1 \left(\frac{1}{2} x^2 - y, \ \frac{1}{y' - x} \right).$$

A more convenient form for this relation is

$$y'' = 1 + g_1 \left(2y - x^2, y' - x \right).$$

For $U_2 f$, we use the same item but with

$$a = 1, \quad m = 1, \quad b = 1, \quad n = 2, \quad r = 0$$

to obtain

$$xyy'' + yy' - 2xy = \frac{y}{x}g_2\left(\frac{1}{2}x^2 - y, \frac{1}{xy' - x^2}\right).$$

This can be written as

$$y'' = 1 - \frac{y' - x}{x} + \frac{1}{x^2}g_2(2y - x^2, x(y' - x)).$$

By choosing

$$g_1 = (y' - x)^2 g\left(2y - x^2\right)$$

and

$$g_2 = x(y' - x) + [x(y' - x)]^2 g\left(2y - x^2\right),$$

the two ODEs become

$$y'' = 1 + (y' - x)^2 g\left(2y - x^2\right). \tag{7.27}$$

This is the ODE that is invariant under the Equations (7.26) symbols with $j = x$. It appears as item IV6 in Table 7.6.

Equation (7.27) is not directly separable. For this equation, we have

$$\rho = j = x, \quad \xi_1 = 1, \quad \eta_1 = x,$$

so that the type IV transformation in Table 7.3 yields

$$X = 2y - x^2, \quad Y = x. \tag{7.28}$$

We thus obtain

$$dX = 2(y' - x)dx$$
$$dY = dx$$
$$Y' = \frac{dY}{dX} = \frac{1}{2(y' - x)}$$
$$Y'' = \frac{dY'}{dX} = -\frac{1}{2}\frac{y'' - 1}{(y' - x)^2}\frac{dx}{dX} = -\frac{1}{4}\frac{y'' - 1}{(y' - x)^3}$$

and consequently Equation (7.27) becomes

$$Y'' = -\frac{1}{2}Y'g(X),$$

in accord with the type IV form in Table 7.2. This relation can be integrated twice, with the result

$$x = a_2 + a_1 \int \exp\left[-\frac{1}{2}\int g(X)dX\right] dX,$$

where X is replaced by $2y - x^2$ after the quadratures are performed, and the a_i are integration constants.

Transformation (7.28) is not unique. For example, the transformation

$$X = 2y - x^2, \quad Y = y' - x$$

also results in a separable form that integrates to the above solution.

7.3   Compendium

Reformulation of the Group Equations

To expedite the later enlargement discussion, two functions of the u_i invariants are introduced

$$p_i = p_i(u_1, u_2), \quad i = 1, 2. \tag{7.29}$$

These provide new symbols

$$\widehat{U}_i = p_i U_i, \quad i = 1, 2, \tag{7.30}$$

where

$$\widehat{\xi}_i = p_i \xi_i, \quad \widehat{\eta}_i = p_i \eta_i, \quad i = 1, 2. \tag{7.31}$$

While it is not necessary, at this time, that the $U_i f$ constitute a two-parameter group, it is required that the $\widehat{U}_i f$ form such a group, i.e.,

$$(\widehat{U}_1 \widehat{U}_2)f = \widehat{e}_1 \widehat{U}_1 f + \widehat{e}_2 \widehat{U}_2 f. \tag{7.32}$$

The first term on the right side of Equation (7.4) becomes

$$\widehat{U}_1 \widehat{\xi}_2 = p_1 U_1 (p_2 \xi_2) = p_1 (\xi_2 U_1 p_2 + p_2 U_1 \xi_2)$$

$$= p_1 \left[\xi_2 \left(\frac{\partial p_2}{\partial u_1} U_1 u_1 + \frac{\partial p_2}{\partial u_2} U_1 u_2 \right) + p_2 U_1 \xi_2 \right]$$

$$= p_1 (\xi_2 p_{2u_2} U_1 u_2 + p_2 U_1 \xi_2)$$

since

$$U_i u_i = 0, \quad i = 1, 2.$$

In a similar manner, we also obtain

$$\widehat{U}_2 \widehat{\xi}_1 = p_2 (\xi_1 p_{1u_1} U_2 u_1 + p_1 U_2 \xi_1)$$

$$\widehat{U}_1 \widehat{\eta}_2 = p_1 (\eta_2 p_{2u_2} U_1 u_2 + p_2 U_1 \eta_2)$$

$$\widehat{U}_2 \widehat{\eta}_1 = p_2 (\eta_1 p_{1u_1} U_2 u_1 + p_1 U_2 \eta_1).$$

Equations (7.4) and (7.32) thereby yield

$$[p_1(\xi_2 p_{2u_2} U_1 u_2 + p_2 U_1 \xi_2) - p_2(\xi_1 p_{1u_1} U_2 u_1 + p_1 U_2 \xi_1)]f_x$$

$$+ [p_1(\eta_2 p_{2u_2} U_1 u_2 + p_2 U_1 \eta_2) - p_2(\eta_1 p_{1u_1} U_2 u_1 + p_1 U_2 \eta_1)]f_y$$

$$= \widehat{e}_1 p_1 (\xi_1 f_x + \eta_1 f_y) + \widehat{e}_2 p_2 (\xi_2 f_x + \eta_2 f_y).$$

By equating the coefficients of f_x and f_y, we have

$$p_1 p_{2u_2} \xi_2 U_1 u_2 - p_2 p_{1u_1} \xi_1 U_2 u_1 + p_1 p_2 (U_1 \xi_2 - U_2 \xi_1) = \widehat{e}_1 p_1 \xi_1 + \widehat{e}_2 p_2 \xi_2$$

$$p_1 p_{2u_2} \eta_2 U_1 u_2 - p_2 p_{1u_1} \eta_1 U_2 u_1 + p_1 p_2 (U_1 \eta_2 - U_2 \eta_1) = \widehat{e}_1 p_1 \eta_1 + \widehat{e}_2 p_2 \eta_2. \tag{7.33}$$

These equations represent a general condition on the p_i for the $\widehat{U}_i f$ to be a two-parameter group. In particular, if the p_i are unity, they simplify to

$$U_1\xi_2 - U_2\xi_1 = e_1\xi_1 + e_2\xi_2$$
$$U_1\eta_2 - U_2\eta_1 = e_1\eta_1 + e_2\eta_2, \tag{7.34}$$

where the $\widehat{e}_i$ have been replaced by e_i. These relations represent the condition for the $U_i f$ to constitute a two-parameter group. Equations (7.33), however, do not require that these relations hold.

The function $\widehat{\phi}$, defined by Equation (7.9), becomes

$$\widehat{\phi} = \widehat{U}_2 f - \widehat{\rho}\widehat{U}_1 f = p_2 U_2 f - \widehat{\rho} p_1 U_1 f$$
$$= (p_1\xi_2 - \widehat{\rho} p_1\xi_1)f_x + (p_2\eta_2 - \widehat{\rho} p_1\eta_1)f_y. \tag{7.35}$$

Again, if the p_i are unity, we have

$$\phi = (\xi_2 - \rho\xi_1)f_x + (\eta_2 - \rho\eta_1)f_y \tag{7.36}$$

with an obvious change in notation. The $\phi = 0$ condition thus yields

$$\rho = \frac{\xi_2}{\xi_1} = \frac{\eta_2}{\eta_1},$$

where, occasionally, one of the ratios may be indeterminant, and $\Delta = 0$.

The substitution principle, see Appendix C, still holds. Thus, the left side of Equations (7.34) transform as

$$U_1\xi_2 - U_2\xi_1 \rightarrow U_1\eta_2 - U_2\eta_1$$

$$U_1\eta_2 - U_2\eta_1 \rightarrow U_1\xi_2 - U_2\xi_1,$$

while on the right side we have

$$\xi_1 \rightarrow \eta_i, \quad \eta_i \rightarrow \xi_i.$$

Preliminary Remarks on Table Generation

The discussion in Sections 7.1 and 7.2 define and characterize two-parameter groups. In contrast to a one-parameter group, a two-parameter group of a specified type must satisfy Equation (7.5) and Table 7.1. The earlier discussion, however, leaves open the question of how such groups are to be generated. One approach would be to select two linearly independent groups from Table 5.4, which generally will not satisfy the two-parameter group requirement. In this circumstance, Equations (7.33) represent two coupled PDEs for p_1 and p_2. Complicating the solution process, however, is the requirement that the coefficients be transformed into functions of the u_1 and u_2 invariants. Even if successful, a transformation to the V_i symbols is then generally required. A discussion of the enlargement procedure is resumed later in this section.

A simpler process based on Equations (7.34) and (7.36), where $p_i = 1$, is adopted. This method enables us to relatively easily generate two-parameter groups of a prescribed type in their fundamental form. The e_i in Equations (7.34) are chosen in accord with Equations (7.6); hence, a transformation to $V_i f$ symbols is unnecessary. Table 7.4 summarizes this approach for each of the four fundamental types. Observe that the left side of the symbol equations is repeated, and that $\xi_i \eta_2 - \xi_2 \eta_1 \neq 0$ insures that $\phi \neq 0$. For purposes of constructing a catalogue of invariant ODEs, the formulation in Table 7.4 is more convenient than that of the preceding paragraph.

The four coefficients, ξ_i and η_i, in Equations (7.34) are *a priori* unknown. Since there are only two equations, two of the coefficients can be chosen arbitrarily. Shortly, we will make such a choice. After this choice is made, Equations (7.34) and (7.36) determine the remaining coefficients to within arbitrary functions of integration.

After the above steps are taken, the symbols are known, i.e., they have specific forms. The two symbols constitute a two-parameter group of a known type with the symbols in their fundamental form. We could then construct the six invariant functions and thereby determine Equations (7.10), after which Equation (7.12) would be found. This approach is to be used in the next section with Equation (7.64).

Finding the invariant functions, however, is not always a simple task. It is simple, however, if both symbols are present in Tables 5.4 and 6.1. In any case, Equations (7.10) are often rather complicated when the various invariant functions are written in terms of x, y, y', and y''. In turn, this complexity can make the determination of Equation (7.12) difficult.

Instead, a simpler procedure is utilized in which *none* of the six invariant functions are evaluated. (The last example in the next section illustrates this approach.) Since the type is known, Table 7.3 is used to directly determine the canonical coordinates, X and Y. With these coordinates, Y' and Y'' are evaluated in terms of x, y, y', and y''. Because the symbols have their fundamental form, the ODE invariant under the group is given by the last column in Table 7.2. Thus, if the group is of type III, the ODE has the separable form

$$XY'' = g(Y').$$

We transform this ODE into one involving x, y, y', and y''. In terms of these variables, the ODE is the desired result; it is still invariant under the group. Although it is usually not separable, it can be made separable by reintroducing canonical coordinates.

Table 7.5

We choose simple values for ξ_1 and η_1, e.g.,

$$\xi_1 = 0, \quad \eta_1 = A(x)B(y) \tag{7.37}$$

for all four types. This choice is but one of many possibilities. It has the advantage of yielding a wide variety of useful ODEs with relatively simple analytical forms. It is the basis of Table 7.5.

With Equations (7.37), the symbol equations for types I and II become (see Table 7.4)

$$\eta_1 \eta_{2y} - \xi_2 \eta_{1x} - \eta_2 \eta_{1y} = 0 \tag{7.38a}$$
$$\xi_{2y} = 0. \tag{7.38b}$$

Equation (7.38b) yields

$$\xi_2 = \gamma(x), \tag{7.39}$$

where γ is an arbitrary function of integration. Equation (7.38a) becomes

$$AB\eta_{2y} - \gamma A' B - \eta_2 AB' = 0,$$

or

$$\eta_{2y} = \frac{B'}{B}\eta_2 + \gamma\frac{A'}{A},$$

where

$$A' = \frac{dA}{dx}, \quad B' = \frac{dB}{dy}.$$

The method of characteristics is used to obtain a solution for η_2, as follows:

$$\frac{dx}{0} = \frac{dy}{1} = \frac{d\eta_2}{(B'/B)\eta_2 + \gamma(A'/A)}$$

$$u = x = c$$

$$\frac{d\eta_2}{dy} - \frac{B'}{B}\eta_2 = \gamma\frac{A'}{A}.$$

This last relation is a linear, first-order ODE whose solution is

$$\eta_2 = B\left(\sigma(x) + \gamma\frac{A'}{A}\int\frac{dy}{B}\right), \tag{7.40}$$

where σ is another arbitrary function of integration.

By way of summary, we see that Equations (7.37) result in Equations (7.39) and (7.40) for the types I and II fundamental forms. In ξ_2 and η_2, γ and σ are functions of x.

For types I and III, the ϕ condition in Table 7.4 results in

$$\gamma AB \neq 0. \tag{7.41}$$

For types II and IV, the ϕ condition yields

$$\rho = \frac{\gamma}{0} = \frac{1}{A}\left(\sigma + \frac{\gamma A'}{A}\int\frac{dy}{B}\right)$$

and consequently

$$\gamma = 0, \quad \rho = \frac{\sigma(x)}{A(x)}. \tag{7.42}$$

From Equation (7.39), we have $\xi_2 = 0$. Table 7.5 summarizes the various $U_2 f$ symbols for all four types.

As previously discussed, Table 7.3 is utilized for the canonical coordinates. For type I, we have

$$ABX_y = 1 \tag{7.43a}$$

$$\gamma X_x + B \left(\sigma + \gamma \frac{A'}{A} \int \frac{dy}{B} \right) X_y = 0 \tag{7.43b}$$

$$ABY_y = 0 \tag{7.43c}$$

$$\gamma Y_x + B \left(\sigma + \gamma \frac{A'}{A} \int \frac{dy}{B} \right) Y_y = 1. \tag{7.43d}$$

Eliminate X_y from Equations (7.43a) and (7.43b), with the result

$$X_x + \frac{\sigma}{\gamma A} + \frac{A'}{A^2} \int \frac{dy}{B} = 0.$$

Integrate with respect to x, to obtain

$$X = \nu(y) - \int \frac{\sigma dx}{\gamma A} + \frac{1}{A} \int \frac{dy}{B},$$

where ν is a function of integration. Substitute this result into Equation (7.43a) to obtain:

$$AB \left(\nu' + \frac{1}{AB} \right) = 1$$

$$\nu' = 0$$

$$\nu = \text{constant} = 0.$$

Hence, $X(x, y)$ is given by

$$X = \frac{1}{A} \int \frac{dy}{B} - \int \frac{\sigma dx}{\gamma A}. \tag{7.44a}$$

Equations (7.43c) and (7.43d) readily yield for $Y(x, y)$

$$Y = \int \frac{dx}{\gamma}. \tag{7.44b}$$

With X and Y known, the Y' derivative is formed as follows:

$$\frac{dX}{dx} = \frac{y'}{AB} - \frac{\sigma}{\gamma A} - \frac{A'}{A^2} \int \frac{dy}{B}$$

$$\frac{dY}{dx} = \frac{1}{\gamma}$$

$$Y' = \frac{dY}{dX} = \frac{dY/dx}{dX/dx} = \frac{AB}{\gamma y' - \sigma B - \gamma \dfrac{A'}{A} B \displaystyle\int \dfrac{dy}{B}}. \tag{7.45a}$$

After some algebraic simplification, the Y'' derivative is given by

$$Y'' = \frac{dY'}{dX} = \frac{dY'/dx}{dX/dx}$$

$$= \frac{\gamma AB}{\left[\gamma y' - \sigma B - (\gamma A' B/A) \displaystyle\int \dfrac{dy}{B}\right]^3} \left\{ -\gamma ABy'' + \gamma AB'(y')^2 + 2\gamma A'By' - \gamma'ABy' \right.$$

$$\left. + \sigma'AB^2 - \sigma A'B^2 + \left[\gamma A'' - \frac{2\gamma(A')^2B^2}{A} + \gamma'A'B^2\right] \int \frac{dy}{B} \right\}. \tag{7.45b}$$

Observe that the coefficient γAB cannot be zero in accord with inequality (7.41).
   In terms of canonical coordinates, the type I ODE has the form

$$Y'' = g(Y'). \tag{7.46}$$

In conjunction with Equations (7.45), this is the general form of the ODE whose symbols are given by Equations (7.37), (7.39), and (7.40). In Equations (7.44) and (7.45), A, γ, and σ are arbitrary functions of x, while B is an arbitrary function of y. The foregoing discussion explains how the type I results in Table 7.5 were arrived at.
   As a simple example of Equation (7.46), suppose

$$A = 1, \quad B = 1.$$

We then have:

$$X = y - \int \frac{\sigma dx}{\gamma}, \quad Y = \int \frac{dx}{\gamma}$$

$$Y' = \frac{1}{\gamma y' - \sigma}$$

$$Y'' = \frac{\gamma(\gamma y'' + \gamma'y' - \sigma')}{(\gamma y' - \sigma)^3}.$$

Equation (7.46) takes the form

$$-\frac{\gamma(\gamma y'' + \gamma'y' - \sigma')}{(\gamma y' - \sigma)^3} = g\left(\frac{1}{\gamma y' - \sigma}\right),$$

which simplifies to

$$\gamma(\gamma y'' + \gamma' y' - \sigma') = g(\gamma y' - \sigma) \tag{7.47}$$

or

$$\gamma(\gamma y' - \sigma)' = g(\gamma y' - \sigma).$$

If we now set

$$\gamma = x, \quad \sigma = 0,$$

the elegant result

$$x^2 y'' = g(x y') \tag{7.48}$$

is obtained as the ODE invariant under the type I group

$$U_1 f = f_y, \quad U_2 f = x f_x. \tag{7.49}$$

Generally, Equation (7.48) is not separable, although it becomes type I separable
with the $X = y$, $Y = \ln x$ transformation.

The substitution principle results in the symbols

$$U_1 f = A(y) B(x) f_x$$
$$U_2 f = B \left(\sigma(y) + \frac{\gamma(y) A'}{A} \int \frac{dx}{B} \right) f_x + \gamma(y) f_y$$

for the type I group in Table 7.5. Similarly, Equations (7.44) and (7.45a) become

$$X = \frac{1}{A} \int \frac{dx}{B} - \int \frac{\sigma dy}{\gamma A}$$
$$Y = \int \frac{dy}{\gamma}$$
$$Y' = \frac{A B y'}{\gamma - B \left(\sigma + \frac{\gamma A'}{A} \int \frac{dx}{B} \right) y'}.$$

In addition, Equations (7.47) and (7.48), respectively, transform to

$$\gamma \left[\gamma y'' - \gamma'(y')^2 + \sigma'(y')^3 \right] = (y')^3 g \left(\frac{\gamma - \sigma y'}{y'} \right) \tag{7.50}$$

and

$$y'' = y' g \left(\frac{y'}{y} \right). \tag{7.51}$$

Since application of the substitution principle is straightforward, results are not
shown for it in the table.

We next briefly discuss some of the Table 7.5 type II results. In accord with Table 7.3 and Equation (7.42), the X canonical coordinate for type II is

$$X = \frac{\sigma(x)}{A(x)} \tag{7.52}$$

for the symbols provided by Equations (7.37), (7.39), and (7.40). For Y, we have

$$A(x)B(y)Y_y = 1,$$

which integrates to

$$Y = \frac{1}{A} \int \frac{dy}{B},$$

where the function of integration is set equal to zero. The type II ODE invariant under the group is

$$Y'' = g(X). \tag{7.53}$$

The derivations for types II, III, and IV are similar to that of type I. Table 7.5 summarizes this material, which can be viewed as a two-parameter counterpart of Table 5.2.

Table 7.6

Table 7.6 contains a more comprehensive catalogue of ODEs invariant under a two-parameter group. The table is largely self-explanatory. The groups are listed by type and the s designation stands for the substitution principle. (Do not confuse the item designation with that used in Tables 5.3, 5.4, and 6.1.) Thus, items I2 and I2s correspond to Equations (7.48) and (7.51), respectively. Many of the items are based on Table 7.5. Canonical coordinates can be derived using Table 7.3. For the many items stemming from Table 7.5, however, this is unnecessary since this table already provides the canonical coordinates. Thus, for common items, both tables can be utilized.

Enlargement Procedure

We presume the symbols, $U_i f, i = 1, 2$, and their enlargement, $\widehat{U}_i f$, are two-parameter groups of the same type and are in their fundamental forms. In view of this, we have

$$\widehat{e}_i = e_i$$

and Equations (7.33) and (7.34) combine to yield

$$\xi_1[p_2 p_{1u_1} U_2 u_1 - p_1 e_1(p_2 - 1)] - \xi_2[p_1 p_{2u_2} U_1 u_2 + p_2 e_2(p_1 - 1)] = 0$$

$$\eta_1[p_2 p_{1u_1} U_2 u_1 - p_1 e_1(p_2 - 1)] - \eta_2[p_1 p_{2u_2} U_1 u_2 + p_2 e_2(p_1 - 1)] = 0.$$

These relations are satisfied for arbitrary ξ_i and η_i if

$$p_2 p_{1u_1} U_2 u_1 - p_1 e_1 (p_2 - 1) = 0$$
$$p_1 p_{2u_2} U_1 u_2 + p_2 e_2 (p_1 - 1) = 0. \tag{7.54}$$

They constitute constraints on the possible forms for the p_i in order that the $\widehat{U}_i f$ represents a group of the same type as $U_i f$.

In particular, for type I and II, we have

$$e_i = 0, \quad i = 1, 2$$

and Equations (7.54) reduce to

$$p_{1u_1} U_2 u_1 = 0, \quad p_{2u_2} U_1 u_2 = 0. \tag{7.55}$$

As a simple example, we find the p_i for item I2 in Table 7.6. We thus obtain the following:

$$U_1 f = f_y, \quad U_2 f = x f_x$$
$$\frac{dx}{0} = \frac{dy}{1} \rightarrow u_1 = x, \quad U_2 u_1 = x$$
$$\frac{dx}{x} = \frac{dy}{0} \rightarrow u_2 = y, \quad U_1 u_2 = 1.$$

Consequently, Equations (7.55) yield

$$p_1(u_2) = p_1(y), \quad p_2(u_1) = p_2(x),$$

and the enlarged group has the symbols

$$\widehat{U}_1 f = p_1(y) f_y, \quad \widehat{U}_2 f = p_2(x) x f_x.$$

This result, however, is equivalent to item I1 in Table 7.6.

It is useful to note that the enlarged group must still satisfy the symbol equations and ϕ condition in Table 7.4. For instance, the first equation for types I and II becomes

$$p_1 \xi_1 (p_2 \eta_2)_x + p_1 \eta_1 (p_2 \eta_2)_y - p_2 \xi_2 (p_1 \eta_1)_x - p_2 \eta_2 (p_1 \eta_1)_y = 0, \tag{7.56}$$

while the types I and III ϕ condition is

$$p_1 \xi_1 p_2 \eta_2 - p_2 \xi_2 p_1 \eta_1 = p_1 p_2 (\xi_1 \eta_2 - \xi_2 \eta_1) \neq 0.$$

The ϕ condition for types II and IV becomes

$$\rho = \frac{p_2 \, \xi_2}{p_1 \, \xi_1} = \frac{p_2 \, \eta_2}{p_1 \, \eta_1}.$$

For item I3 in Table 7.6, we obtain

$$u_1 = x, \quad U_2 u_1 = 0$$
$$u_2 = x, \quad U_1 u_2 = 0,$$

and Equations (7.55) no longer constrain the p_i. In this case, an equation such as Equation (7.56) need to be satisfied.

As a final example, we consider item I6 in Table 7.6. From Table 7.2, the invariant function u_1 is obtained

$$u_1 = \frac{y}{x}.$$

For u_2, we have

$$\frac{dx}{x+y} = \frac{dy}{y-x},$$

which is integrated in the next section [see Equation (7.68)], with the result

$$u_2 = (x^2 + y^2)^{1/2} \exp[\tan^{-1}(y/x)].$$

Because of the complexity of u_2, we set $p_1 = 1$, while for p_2 we have

$$p_{2u_2} = 0$$

or

$$p_2 = p_2(u_1) = p_2(y/x). \tag{7.57}$$

The enlarged group thus has the form

$$\widehat{U}_1 f = U_1 f, \quad \widehat{U}_2 f = p_2\left(\frac{y}{x}\right)[(x+y)f_x + (y-x)f_y].$$

The equations for the canonical coordinates of the enlarged group are:

$$xX_x + yX_y = 1 \tag{7.58a}$$
$$p_2(x+y)X_x + p_2(y-x)X_y = 0 \tag{7.58b}$$
$$xY_x + yY_y = 0 \tag{7.58c}$$
$$p_2(x+y)Y_x + p_2(y-x)Y_y = 1, \tag{7.58d}$$

where p_2 cancels in Equation (7.58b). As shown in the next section [see Equations (7.73)], Equations (7.58a) and (7.58b) can be integrated. Integration of the Y equations results in

$$Y = -\int \frac{d(y/x)}{[1 + (y/x)^2]p_2(y/x)}.$$

See Problem 7.15 for the ODE invariant under the enlarged group.

In summary, the enlargement procedure can be extended to two-parameter groups, but it lacks the flexibility associated with the one-parameter groups.

7.4   Examples

As a first example, we find the general solution to

$$y'' + ax^n y' = (y')^2 g(y),$$ (7.59)

where a and n are constants and $g(y)$ is an arbitrary function. This ODE conforms to item IV1s in Table 7.6, which is a special case of type IV in Table 7.5. For $B(x)$, we have

$$\frac{B'}{B} = ax^n,$$

which integrates to

$$B = \exp\left(\frac{a}{n+1}x^{n+1}\right), \quad n \ne -1.$$

Table 7.5 is utilized with this B and

$$\sigma = 0, \quad A = 1$$

to obtain, by means of the substitution principle,

$$X = y, \quad Y = \int \exp\left(-\frac{a}{n+1}x^{n+1}\right) dx.$$ (7.60)

The type IV ODE

$$Y'' = Y'g(X)$$

integrates to

$$\frac{dY'}{Y'} = g(X)dX$$

$$\ln Y' = \text{constant} + \int g dX$$

$$Y' = b_1 \exp\left(\int g dX\right).$$

After a second integration, we have the general solution of Equation (7.59)

$$Y = b_2 + b_1 \int \exp\left(\int g(X)dX\right) dX,$$ (7.61)

where the b_i are constants of integration and g is the negative of the same function as in Equation (7.59). After the two quadratures are performed, the canonical variables are replaced with x and y. For Y, this requires a third quadrature.

The next example considers the ODE

$$yy'' + \lambda(y')^2 + \beta y^2 = 0,$$ (7.62)

where λ and δ are constants. This is a neutron diffusion equation that arises in nuclear reactor theory (Axford, 1971). The ODE is rewritten as

$$y'' = -\frac{\lambda}{y}(y')^2 - \beta y = y'\left(-\lambda\frac{y'}{y} - \beta\frac{y}{y'}\right),$$

and we see that this has the form of item I2s in Table 7.6 with

$$g = -\lambda\frac{y'}{y} - \beta\frac{y}{y'}.$$

From Tables 7.5 and 7.6, we have

$$A = B = 1, \quad \gamma = y, \quad \sigma = 0$$

$$X = \frac{1}{A}\int\frac{dx}{B} = x, \quad Y = \int\frac{dy}{\gamma} = \ln y. \tag{7.63}$$

With the aid of Table 7.5,

$$Y' = \frac{y'}{y}, \quad Y'' = \frac{y'}{y}\frac{y''}{y'} - \left(\frac{y'}{y}\right)^2.$$

Equation (7.62) becomes

$$\frac{d^2Y}{dX^2} = -\lambda\left(\frac{dY}{dX}\right)^2 - \beta - \left(\frac{dY}{dX}\right)^2$$

$$= -(1+\lambda)\left(\frac{dY}{dX}\right)^2 - \beta.$$

For a first integration, set

$$Z = \frac{dY}{dX},$$

thereby yielding

$$\frac{dZ}{dX} = -(1+\lambda)Z^2 - \beta$$

$$-X = -a + \frac{1}{1+\lambda}\int\frac{dZ}{Z^2 + \beta(1+\lambda)^{-1}}$$

$$X = a - \frac{1}{1+\lambda}\left(\frac{1+\lambda}{\beta}\right)^{1/2}\tan^{-1}\left[\left(\frac{1+\lambda}{\beta}\right)^{1/2}Z\right]$$

$$= a - \frac{1}{[\beta(1+\lambda)]^{1/2}}\tan^{-1}\left[\left(\frac{1+\lambda}{\beta}\right)^{1/2}\frac{dY}{dX}\right],$$

where a is an integration constant. We next solve for dY/dX

$$\left(\frac{1+\lambda}{\beta}\right)^{1/2}\frac{dY}{dX} = -\tan\{[\beta(1+\lambda)]^{1/2}X + \tilde{a}\},$$

where

$$\tilde{a} = -[\beta(1+\lambda)]^{1/2}a.$$

We now proceed with a second integration

$$\left(\frac{1+\lambda}{\beta}\right)^{1/2} Y = \left(\frac{1+\lambda}{\beta}\right)^{1/2} \ln b - \frac{1}{[\beta(1+\lambda)]^{1/2}} \int \tan \tilde{Z}\, d\tilde{Z},$$

where

$$\tilde{Z} = [\beta(1+\lambda)]^{1/2}X + \tilde{a}$$

and

$$Y = \ln b - \frac{1}{1+\lambda}(-\ln b \cos \tilde{Z})$$

$$= \ln b + \frac{1}{1+\lambda}\ln(\cos\{[\beta(1+\lambda)]^{1/2}X + \tilde{a}\}).$$

Transforming back to the original variables then yields

$$\ln y = \ln b + \ln(\cos\{[\beta(1+\lambda)]^{1/2}x + \tilde{a}\})^{1/(1+\lambda)}$$

or

$$y = b(\cos\{[\beta(1+\lambda)]^{1/2}(x-a)\})^{1/(1+\lambda)}$$

for the general solution of Equation (7.62), where a and b are integration constants.

The final two examples are similar in that the symbols are provided and we are to find the type, canonical coordinates, and second-order ODE invariant under the group. These examples thus illustrate how a table can be constructed. They are further chosen because of the variety of techniques utilized for finding the invariant ODE.

In the first of these examples, we have

$$U_1 f = x f_x + y f_y, \quad U_2 f = (x+y) f_x + (y-x) f_y. \qquad (7.64)$$

We start with

$$\xi_1 = x, \quad \eta_1 = y, \quad \xi_2 = x+y, \quad \eta_2 = y-x$$

and determine

$$U_1 \xi_2 = \left(x\frac{\partial}{\partial x} + y\frac{\partial}{\partial y}\right)(x+y) = x+y$$

$$U_2 \eta_1 = x+y$$

$$U_1 \eta_2 = y-x$$

$$U_2 \eta_1 = y-x.$$

Hence, Equation (7.4) yields

$$(U_1 U_2)f = 0,$$

while

$$\phi = -\rho(x f_x + y f_y) + (x + y) f_x + (y - x) f_y \neq 0.$$

Thus, the group is of type I.

We next utilize item IIIa3 in Tables 5.4 and 6.1, with $a = b = 1$, to obtain

$$x^2 y'' = y g_1 \left(\frac{x}{y}, \frac{x}{y} y' \right)$$

or

$$G_1(y/x, y', x y'') = 0$$

for the $U_1 f$ symbol. In view of the equivalence concept, observe that G_1 can also be written as

$$G_1 \left(\frac{y}{x}, \frac{(y/x) - y'}{1 + (y/x) y'}, \frac{[1 + (y/x)^2]^2 x y''}{[(y/x) - y']^3} \right) = 0,$$

which simplifies to

$$G_1 \left[\frac{y}{x}, \frac{y - x y'}{x + y y'}, \frac{(x^2 + y^2)^2 y''}{(y - x y')^3} \right] = 0. \qquad (7.65)$$

This relation will be compared to G_2, which we now derive.

For the $U_2 f$ symbol, we have

$$\eta_2' = -1 - (y')^2$$
$$\eta_2'' = -(1 + 3y') y''$$

and Equations (6.4) become

$$\frac{dx}{x + y} = \frac{dy}{y - x} = -\frac{dy'}{1 + (y')^2} = -\frac{dy''}{(1 + 3y') y''}. \qquad (7.66)$$

For the leftmost homogeneous equation

$$\frac{dy}{dx} = \frac{y - x}{y + x}$$

set

$$y = x\nu$$
$$dy = x\, d\nu + \nu\, dx \qquad (7.67)$$

with the result

$$\frac{1 + \nu}{1 + \nu^2} d\nu = -\frac{dx}{x}.$$

This integrates to

$$\ln x + \ln(1 + \nu^2)^{1/2} + \tan^{-1} \nu = \ln c$$

or

$$u_2 = (x^2 + y^2)^{1/2} \exp[\tan^{-1}(y/x)] = c. \tag{7.68}$$

This relation cannot be solved explicitly for either x or y, thereby compromising our ability to find $u_2^{(1)}$ and $u_2^{(2)}$. Furthermore, the Riccati formulation in Section 5.1 is of no help in this circumstance.

This difficulty is overcome by retaining ν as a variable. Equation (7.68) thus becomes

$$x = \frac{c}{(1+\nu^2)^{1/2}} \exp(-\tan^{-1}\nu). \tag{7.69}$$

We shall also need the derivative

$$\frac{dx}{d\nu} = -c\frac{1+\nu}{(1+\nu^2)^{3/2}} \exp(-\tan^{-1}\nu).$$

Hence, the leftmost factor in Equations (7.66) equals

$$\frac{dx}{x+y} = -\frac{d\nu}{1+\nu^2}$$

after simplification. The first differential invariant thereby stems from

$$\frac{d\nu}{1+\nu^2} = \frac{dy'}{1+(y')^2},$$

which integrates to

$$\tan^{-1}\nu = \tan^{-1}y' + \tilde{c}_1.$$

With

$$\theta = \tan^{-1}\nu, \quad \psi = \tan^{-1}y',$$

we obtain

$$\tan(\theta - \psi) = \frac{\tan\theta - \tan\psi}{1 + \tan\theta\tan\psi} = \frac{\nu - y'}{1 + \nu y'} = \frac{y - xy'}{x + yy'}.$$

Consequently, the first differential invariant is

$$u_2^{(1)} = \frac{y - xy'}{x + yy'} = c_1. \tag{7.70}$$

For the second differential invariant, we use

$$\frac{d\nu}{1+\nu^2} = \frac{dy''}{(1+3y')y''}.$$

From Equations (7.67) and (7.70), we have

$$1 + 3y' = 1 + \frac{3(\nu - c_1)}{1 + c_1\nu},$$

with the result

$$\frac{dy''}{y''} = \frac{d\nu}{1+\nu^2} + \frac{3(\nu - c_1)d\nu}{(1+c_1\nu)(1+\nu^2)}.$$

[Note that it is simpler to use the dy', dy'' equation in Equations (7.66).] The rightmost term is integrated by partial fractions, i.e.,

$$\frac{3(\nu - c_1)d\nu}{(1+c_1\nu)(1+\nu^2)} = -\frac{3c_1 d\nu}{1+c_1\nu} + \frac{3\nu d\nu}{1+\nu^2}.$$

We thereby obtain

$$\ln y'' = \ln c_2 + \tan^{-1}\nu - 3\ln\left(\nu + \frac{1}{c_1}\right) + \frac{3}{2}\ln\left(1 + \nu^2\right).$$

Next, eliminate c_1 and ν and simplify, with the result

$$u_2^{(2)} = \frac{(x^2+y^2)^{3/2}y''}{(y-xy')^3}\exp[-\tan^{-1}(y/x)] = c_2. \tag{7.71}$$

The exponential factor is eliminated using Equation (7.68), and rearrange $u_2^{(2)}$ as follows:

$$u_2^{(2)} = \frac{(x^2+y^2)^2 y''}{u_2(y-xy')^3}.$$

Thus, G_2 has the form

$$G_2\left((x^2+y^2)^{1/2}\exp\left[\tan^{-1}(y/x)\right], \frac{y-xy'}{x+yy'}, \frac{(x^2+y^2)^2 y''}{(y-xy')^3}\right) = 0.$$

By comparing G_2 with G_1, we see that the ODE invariant under the group has the form

$$G\left[\frac{y-xy'}{x+yy'}, \frac{(x^2+y^2)^2 y''}{(y-xy')^3}\right] = 0, \tag{7.72}$$

which is in accord with item I6 in Table 7.6.

Since the ODE is type I, the transformation equations are

$$xX_x + yX_y = 1$$
$$(x+y)X_x + (y-x)X_y = 0$$
$$xY_x + yY_y = 0 \tag{7.73}$$
$$(x+y)Y_x + (y-x)Y_y = 1.$$

The first two equations are solved for X_y, with the result

$$X_y = \frac{x+y}{x^2+y^2},$$

which integrates to

$$X = \nu(x) + \tan^{-1}(y/x) + \ln\left(x^2+y^2\right)^{1/2}, \tag{7.74a}$$

where ν is a function of integration. From the second of Equations (7.73), we conclude that ν can be set equal to zero. A similar procedure for Y yields

$$Y = -\tan^{-1}(y/x). \tag{7.74b}$$

Equations (7.74), with $\nu = 0$, provide the canonical coordinates.

The symbols for the final example are

$$U_1 f = \frac{x}{x+y} f_x + \frac{y}{x+y} f_y$$

$$U_2 f = \frac{x^2}{x+y} f_x + \frac{y^2 + 2xy}{x+y} f_y. \tag{7.75}$$

We now obtain

$$U_1 \xi_2 = \frac{x^2}{(x+y)^2}$$

$$U_2 \xi_1 = -\frac{xy}{(x+y)^2}$$

$$U_1 \eta_2 = \frac{y(2x+y)}{(x+y)^2}$$

$$U_2 \eta_1 = \frac{xy}{(x+y)^2},$$

which results in the commutator

$$(U_1 U_2) f = \frac{x}{x+y} f_x + \frac{y}{x+y} f_y = U_1 f.$$

The function ϕ can be written as

$$\phi = \frac{x}{x+y}(x - \rho) f_x + \frac{y}{x+y}(y + 2x - \rho) f_y \neq 0.$$

Hence, the group is of type III. It is not listed in Tables 7.5 or 7.6.

As noted in Section 7.2, it is not essential that the invariants be determined, as was done in the preceding example. Instead, the canonical variables are directly obtained from the transformation equations

$$x X_x + y X_y = 0 \tag{7.76a}$$

$$x^2 X_x + (y^2 + 2xy) X_y = (x + y) X \tag{7.76b}$$

$$x Y_x + y Y_y = x + y \tag{7.76c}$$

$$x^2 Y_x + (y^2 + 2xy) Y_y = (x + y) Y. \tag{7.76d}$$

Multiply Equation (7.76a) by x and subtract the result from the second equation to obtain

$$y X_y - X = 0$$

$$\frac{\partial}{\partial y}\left(\frac{X}{y}\right) = 0$$

$$X = y\nu(x),$$

where ν is a function of integration. Substitute this relation into Equation (7.76a) with the result

$$x\nu' + \nu = 0.$$

A particular solution is $\nu = x^{-1}$, which yields

$$X = \frac{y}{x}. \qquad (7.77a)$$

The same procedure with regard to Equations (7.76c) and (7.76d) results in

$$yY_y - Y = -x$$

$$y^2 \frac{\partial}{\partial y} \left(\frac{Y}{y} \right) = -x$$

$$Y = x + y\mu(x),$$

where μ is a function of integration. Substitute this into Equation (7.76c) to obtain

$$x\mu' + \mu = 1.$$

A particular solution, $\mu = 1$, results in

$$Y = x + y. \qquad (7.77b)$$

Equations (7.77) provide the canonical coordinates.

These equations are differentiated, with the result

$$dX = \frac{xy' - y}{x^2} dx$$

$$dY = (1 + y')dx$$

$$Y' = \frac{x^2(1 + y')}{xy' - y}.$$

By means of a second differentiation, we obtain

$$Y'' = \frac{dx}{dX} \left[\frac{2x(1 + y')}{xy' - y} + \frac{x^2 y''}{xy' - y} - \frac{x^3(1 + y')y''}{(xy' - y)^2} \right],$$

which simplifies to

$$Y'' = \frac{2x^3(1 + y')}{(xy' - y)^2} - \frac{x^4(x + y)y''}{(xy' - y)^3}.$$

For a type III ODE, we have

$$XY'' = g(Y').$$

This becomes

$$\frac{x}{y} \left[-\frac{x^4(x + y)y''}{(xy' - y)^3} + \frac{2x^3(1 + y')}{(xy' - y)^2} \right] = g \left[\frac{x^2(1 + y')}{xy' - y} \right]$$

or, after simplification,

$$\frac{x(x+y)}{(1+y')(xy'-y)}y'' = 2 + \left(\frac{xy'-y}{y}\right)g\left[\frac{x^2(1+y')}{xy'-y}\right] \tag{7.78}$$

for the ODE invariant under the symbols of Equations (7.75).

One limitation of the two-parameter group method is the complexity of the resulting differential equation, e.g., Equations (7.72) and (7.78). It is not likely that either of these equations will arise in scientific or engineering practice. Even if a specific version of either of these equations did occur, it might be fortuitous if they were recognized as special cases. Of course, not all the invariant ODEs are this complicated, as shown by many of the ODEs in Table 7.6. Moreover, as Equation (7.62) illustrates, some of the ODEs in the table do occur in practice.

7.5   Problems

7.1. Solve the ODE

$$y'' = y' + \frac{1}{y}(y')^2.$$

7.2. Solve the ODE

$$xyy'' = 2x(y')^2 + 2yy' + \frac{2}{x}y^2 + x^2y^3.$$

7.3. Solve the ODE

$$yy'' + a\left[1 + (y')^2\right] = 0,$$

where a is a constant. Carry a as a free parameter as long as possible, then solve for $a = \frac{1}{2}$.

7.4. Use Table 7.6 to solve

$$yy'' = (y')^2 + (y'-y)^2 ye^{-x}.$$

7.5. Start with the symbols

$$U_1f = f_y, \quad U_2f = xf_x$$

and derive the second-order ODE invariant under the group and the canonical coordinates.

7.6. Find the type, canonical coordinates, and ODE invariant under the group

$$U_1f = x(x+y)f_x + y(x+y)f_y$$
$$U_2f = x(x-y)f_x + y(x-y)f_y.$$

Be sure to simplify all results.

7.7. Repeat Problem 7.6 for

$$U_1 f = x f_x + y f_y$$
$$U_2 f = x^2 f_x + (y^2 + 2xy) f_y.$$

7.8. Derive the relation

$$(V_1 V_2) f = (\epsilon_{11} \epsilon_{22} - \epsilon_{12} \epsilon_{21})(U_1 U_2) f$$

given after Equations (7.7).

7.9. Repeat Problem 7.6 for

$$U_1 f = x f_y$$
$$U_2 f = x^2 f_x + (y + xy) f_y.$$

7.10. Repeat Problem 7.6 for

$$U_1 f = x^2 f_x - y^2 f_y$$
$$U_2 f = \frac{x^2}{y} f_x - y f_y.$$

7.11. Repeat Problem 7.6 for

$$U_1 f = y f_x - x f_y$$
$$U_2 f = \left(x^2 + y^2\right)^{1/2} y f_x - \left(x^2 + y^2\right)^{1/2} x f_y$$

and determine the ODE by using the substitution principle.

7.12. Solve the ODE

$$y^3 (y'' + 1) = x y' \left[x^2 (y')^2 - 3xyy' + 3y^2 \right].$$

7.13. Use the theory of this chapter to find the general solution to

$$y'' + a e^y = 0,$$

where a is a constant.

7.14. Find the type, canonical coordinates, and ODE invariant under the group

$$U_1 f = x f_x - 2 f_y$$
$$U_2 f = x (\ln x) f_x - 2(1 + \ln x) f_y.$$

Then find the general solution to

$$xy'' + y' + axe^y = 0,$$

where a is a constant.

7.15. With

$$p_2 = -\frac{(y/x)^{1-n}}{n[1 + (y/x)^2]}, \quad n \neq 0,$$

determine the second-order ODE invariant under the enlarged group whose canonical coordinates are provided by Equations (7.58).

Part III

Appendices

Appendix A

Bibliography and References

Ames, W. F., *Nonlinear Ordinary Differential Equations in Transport Processes*, Academic Press, New York, 1968.

Axford, R. A., "Differential Equations Invariant Under Two-Parameter Lie Groups with Applications to Nonlinear Diffusion," Los Alamos Scientific Laboratory, LA-4517, March 1971.

Bluman, G. W. and Cole, J. D., *Similarity Methods for Differential Equations*, Applied Mathematical Sciences, Vol. 13, Springer-Verlag, New York, 1974.

Bluman, G. W. and Kumei, S., *Symmetries and Differential Equations*, Applied Mathematical Sciences, Vol. 81, Springer-Verlag, New York, 1989.

Chester, W., "Continuous Transformations and Differential Equations, *J. Institute Mathematics and Applications*, Vol. 19, 1977, pp. 343–376.

Cohen, A., *An Introduction to the Lie Theory of One-Parameter Groups*, Heath Co., 1911; republished (with errata sheet) by Stechert & Co., New York, 1931.

Dickson, L. E., "Differential Equations from the Group Standpoint," *Annals of Mathematics*, Second Series, Vol. 25, 1923-24, pp. 287–378.

Emanuel, G., *Gasdynamics: Theory and Applications*, American Institute of Aeronautics and Astronautics, Educational Series, Washington, D. C., 1986.

Franklin, P., "The Canonical Form of a One-Parameter Group," *Annals of Mathematics*, Series 2, Vol. 29, 1928, pp. 113–122.

Gilmore, R., *Lie Groups, Lie Algebras, and Some of Their Applications*, John Wiley & Sons, New York, 1974.

Hill, J. M., *Solutions of Differential Equations by Means of One-Parameter Groups*, Pitman Advanced Publishing Program, Boston, 1982.

Ibragimov, N. H., *CRC Handbook of Lie Group Analysis of Differential Equations*, Vol. 1, CRC Press, Boca Raton, FL, 1994.

Ince, E. L., *Ordinary Differential Equations*, Dover Publications, New York, 1956.

Latter, R., "Temperature Behavior of the Thomas-Fermi Statistical Model for Atoms," *Physics Review*, Vol. 99, Sept. 1955, pp. 1854–1870.

Ovsiannikov, L. V., *Group Analysis of Differential Equations*, Academic Press, New York, 1983.

Polyanin, A. D. and Zaitsev, V. F., *Handbook of Exact Solutions for Ordinary Differential Equations*, CRC Press, Boca Raton, FL, 1995.

Rainville, E. D., *Intermediate Course in Differential Equations*, John Wiley & Sons, London, 1943.

Rodriguez Azara, J. L., "A MAPLE program for the generation of the Lie-series solution of systems of non-linear ordinary differential equations," *Comp. Phys. Comm.*, Vol. 67, 1992, pp. 537–542.

Schwarz, F., "Symmetries of Differential Equations: From Sophus Lie to Computer Algebra," *Siam Review*, Vol. 30, 1988, pp. 450–481.

Appendix B

The Rotation Group

1. Global equations:

$$x_1 = x \cos \alpha - y \sin \alpha$$
$$y_1 = \sin \alpha + y \cos \alpha,$$

where α is the group parameter and $\alpha_o = 0,\ \overline{\alpha} = \alpha$.

2. Infinitesimal transformation:

$$\xi = -y, \quad \eta = x$$
$$\eta' = 1 + (y')^2$$
$$\eta'' = 3y'y''$$
$$Uf = -yf_x + xf_y$$
$$U^{(1)}f = Uf + \left[1 + (y')^2\right] f_{y'}$$
$$U^{(2)}f = U^{(1)}f + 3y'y'' f_{y''}.$$

3. First-order invariant ODE:

$$G\left(x^2 + y^2, \frac{y - xy'}{x + yy'}\right) = 0,$$

or

$$y' = \frac{y - xg(x^2 + y^2)}{x + yg(x^2 + y^2)}.$$

With canonical variables

$$X = x^2 + y^2, \quad Y = \tan^{-1}(y/x)$$

we have

$$\frac{dY}{dX} = \frac{1}{X} g(X).$$

147

4. Second-order invariant ODE:

$$G\left(x^2 + y^2, \frac{y - xy'}{x + yy'}, \frac{y''}{[1 + (y')^2]^{3/2}}\right) = 0,$$

or

$$y'' = [1 + (y')^2]^{3/2} g\left(x^2 + y^2, \frac{y - xy'}{x + yy'}\right),$$

or

$$\frac{d\tilde{Y}}{dX} = -\frac{1 + \tilde{Y}^2}{2X}\left\{\tilde{Y} + [X(1 + \tilde{Y}^2)]^{1/2} H(X, \tilde{Y})\right\},$$

where H is arbitrary and

$$\tilde{Y} = \frac{y - xy'}{x + yy'}.$$

Appendix C

Basic Relations

1. Global equations:

$$x_1 = \phi(x, y, \alpha), \quad y_1 = \psi(x, y, \alpha) \tag{2.3}$$

2. Symbols

$$\xi = \left(\frac{\partial \phi}{\partial \alpha}\right)_o, \quad \eta = \left(\frac{\partial \psi}{\partial \alpha}\right)_o \tag{2.8}$$

$$Uf = \xi f_x + \eta f_y \tag{2.12}$$

$$U^{(1)}f = Uf + \eta' f_{y'} \tag{4.15}$$

$$U^{(2)}f = U^{(1)}f + \eta'' f_{y''} \tag{4.17}$$

$$\eta' = \frac{d\eta}{dx} - y'\frac{d\xi}{dx} = \eta_x + (\eta_y - \xi_x)y' - \xi_y(y')^2 \tag{4.14b}$$

$$\eta'' = \frac{d\eta'}{dx} - y''\frac{d\xi}{dx} = \eta_{xx} + (2\eta_{xy} - \xi_{xx})y' + (\eta_{yy} - 2\xi_{xy})(y')^2$$
$$-\xi_{yy}(y')^3 + (\eta_y - 2\xi_x - 3\xi_y y')y'' \tag{4.19}$$

$$\vdots$$

$$\eta^{(n)} = \frac{d\eta^{(n-1)}}{dx} - y^{(n)}\frac{d\xi}{dx} \tag{4.22}$$

3. Lie series:

$$x_1 = x + \alpha\xi + \frac{\alpha^2}{2!}U\xi + \cdots$$

$$y_1 = y + \alpha\eta + \frac{\alpha^2}{2!}U\eta + \cdots \tag{2.18}$$

149

or

$$x_1 = \sum_{n=0}^{\infty} \frac{a^n}{n!} U^n x$$

$$y_1 = \sum_{n=0}^{\infty} \frac{a^n}{n!} U^n y \qquad (2.17)$$

4. Riccati equation solution for $u^{(1)}$:

(a)

$$\frac{dx}{\xi} = \frac{dy'}{\eta'} \qquad (5.5a)$$

$$y = w(x, c) \qquad (3.20a)$$

$$P(x) = \frac{\eta_y - \xi_x}{\xi} - \frac{2\xi_y \eta}{\xi^2}, \quad Q(x) = \frac{\xi_y}{\xi} \qquad (5.12)$$

$$u_{Rx}^{(1)} = \frac{\xi \exp\left(\int P dx\right)}{\xi y' - \eta} - \int Q \exp\left(\int P dx\right) dx \qquad (5.14)$$

(b)

$$\frac{dy}{\eta} = \frac{dy'}{\eta'} \qquad (5.5b)$$

$$x = w(y, c) \qquad (3.20b)$$

$$P(y) = \frac{\xi_x - \eta_y}{\eta} - \frac{2\xi \eta_x}{\eta^2}, \quad Q(y) = \frac{\eta_x}{\eta}$$

$$u_{Ry}^{(1)} = \frac{\eta y' \exp\left(\int P dy\right)}{\eta - \xi y'} - \int Q \exp\left(\int P dy\right) dy$$

5. Canonical coordinates for a first-order ODE:

$$X = u(x, y) \qquad (5.27)$$

$$Y = \int \frac{dy}{\eta}, \quad Y = \int \frac{dx}{\xi} \qquad (5.29)$$

This yields

$$\frac{dY}{dX} = g(X). \qquad (5.26)$$

Other forms are also available.

6. Second-order equation:

$$\frac{d^2Y}{dX^2} = \frac{dY'}{dx}\frac{dx}{dX} = \frac{Y'_x + Y'_y y' + Y'_{y'} y''}{X_x + X_y y'} \tag{6.50}$$

$$\frac{d^2Y}{dX^2} = g\left(X, \frac{dY}{dX}\right) \tag{6.23}$$

7. Substitution principle:

$$\xi \rightleftharpoons \eta$$
$$x \rightleftharpoons y \tag{5.36}$$
$$y' \rightleftharpoons 1/y'$$

$$y'' \rightleftharpoons -\frac{y''}{(y')^3} \tag{6.45}$$

8. Enlargement procedure:

$$\widehat{U}f = h(x, y)Uf \tag{5.43}$$

$$\widehat{P}(x) = P(x) - \frac{1}{\xi h}Uh, \quad \widehat{Q}(x) = Q(x) + \frac{h_y}{h}$$

$$\widehat{u}_{Rx}^{(1)} = \frac{\xi \exp\left(\int \widehat{P}dx\right)}{\xi y' - \eta} - \int \widehat{Q}\exp\left(\int \widehat{P}dx\right)dx \tag{5.44}$$

If $h = p(u)$, then

$$\frac{dY}{dX} + q(X)Y = g(X), \tag{5.50}$$

where X and Y are given by Equations (5.27) and (5.29) and

$$Y = \exp\left(-\int qdX\right)\left[c_1 + \int g\exp\left(\int qdX\right)dX\right]. \tag{5.53a}$$

By the X, Y interchange procedure,

$$\frac{dX}{dY} + Xq(Y) = g(Y), \tag{5.52}$$

where X and Y are given by Equations (5.27) and (5.29) and

$$X = \exp\left(-\int qdY\right)\left[c_1 + \int g\exp\left(\int qdY\right)dY\right]. \tag{5.53b}$$

9. Invariants:

$$u = u(x, y) \qquad\qquad (2.22\text{b})$$

$$u^{(1)} = u^{(1)}(x, y, y') \qquad\qquad (5.6)$$

$$u^{(2)} = u^{(2)}(x, y, y', y'') = \frac{du^{(1)}/dx}{du/dx}$$

$$u^{(3)} = u^{(3)}(x, y, y', y'', y''') = \frac{du^{(2)}/dx}{du/dx}$$

$$\vdots$$

10. Two-parameter formulas:

$$(U_1 U_2)f = U_1(U_2 f) - U_2(U_1 f) \qquad\qquad (7.1)$$
$$= (U_1 \xi_2 - U_2 \xi_1)f_x + (U_1 \eta_2 - U_2 \eta_1)f_y \qquad\qquad (7.4)$$
$$(U_1 U_2)f = e_1 U_1 f + e_2 U_2 f \qquad\qquad (7.5)$$
$$\phi = U_2 f - \rho(x, y)U_1 f \qquad\qquad (7.9)$$

See Table 7.1 for the four fundamental forms.

See Table 7.2 for the canonical (separable) forms of a second-order ODE.

See Table 7.3 for the transformation to canonical coordinates.

See Table 7.4 for constraints for a two-parameter group to be in its fundamental form.

Appendix D

Tables

Table 2.1 Elementary One-Parameter Continuous Groups

Group	Global Equations	Symbol
Translation parallel to x-axis	$x_1 = x + \alpha$ $y_1 = y$	$Uf = f_x$
Translation parallel to y-axis	$x_1 = x$ $y_1 = y + \alpha$	$Uf = f_y$
Uniform radial dilatation	$x_1 = e^\alpha x$ $y_1 = e^\alpha y$	$Uf = xf_x + yf_y$
Rotation about the origin	$x_1 = x \cos \alpha - y \sin \alpha$ $y_1 = x \sin \alpha + y \cos \alpha$	$Uf = -yf_x + xf_y$

Table 5.1. Selected First-Order ODEs Invariant Under a One-Parameter Group

Item	Uf	Group		ODE
1	f_y	Translation paralleled to y-axis	$G(x, y') = 0$	$y' = g(x)$
2	f_x	Translation parallel to x-axis	$G(y, y') = 0$	$y' = g(y)$
3	$\phi(x)f_x$	Affine transformation, $x_1 = x + \alpha\phi(x)$, $y_1 = y$	$G(y, \phi(x)y') = 0$	$\phi(x)y' = g(y)$
4	$\psi(y)f_y$	Affine transformation, $x_1 = x$, $y_1 = y + \alpha\psi(y)$	$G(x, y'/\psi(y)) = 0$	$y' = \psi(y)g(x)$
5	$xf_x + yf_y$	Uniform magnifcation, $x_1 = e^\alpha x$, $y_1 = e^\alpha y$	$G(y/x, y') = 0$	$y' = g(y/x)$
6	$-yf_x + xf_y$	Rotation about the origin	$G\left(x^2 + y^2, \dfrac{y - xy'}{x + yy'}\right) = 0$	$y' = \dfrac{y - xg(x^2 + y^2)}{x + yg(x^2 + y^2)}$

Table 5.2. General Symbols, Canonical Coordinates, and ODE whose Canonical Coordinate Form is $dY/dX = g(X)$

Type	Uf	ODE	Canonical Coordinates
I	$A(x)B(y)f_x$	$G\left(y, \dfrac{1}{Ay'} - \dfrac{B'}{B}\displaystyle\int\dfrac{dx}{A}\right) = 0$	$X = y,\ Y = \dfrac{1}{B}\displaystyle\int\dfrac{dx}{A}$
II	$A(x)f_x + B(x)C(y)f_y$	$G\left(\displaystyle\int\dfrac{B}{A}dx - \int\dfrac{dy}{C},\ \dfrac{A}{C}y' - B\right) = 0$	$X = \displaystyle\int\dfrac{B}{A}dx - \int\dfrac{dy}{C},\ Y = \int\dfrac{dx}{A}$
III	$\dfrac{P_y f_x - P_x f_y}{P_y Q_x - P_x Q_y}E(Q)$	$G\left(P,\ \dfrac{1}{E}\dfrac{Q_x + Q_y y'}{P_x + P_y y'}\right) = 0$	$X = P(x,y),\ Y = \displaystyle\int\dfrac{dQ}{E(Q)}$

Table 5.3. Index for First-Order ODEs

Type	η	ξ			
		(a)	(b)	(c)	(d)
I	0	$M(x)$	$N(y)$	$M(x)N(y)$	$\xi(x,y)$
II	$A(x)$	$M(x)$	$N(y)$	$M(x)N(y)$	$\xi(x,y)$
III	$B(y)$	$M(x)$	—	$M(x)N(y)$	$\xi(x,y)$
IV	$A(x)B(y)$	—	—	$M(x)N(y)$	$\xi(x,y)$
V	$\eta(x,y)$	$\longleftarrow$	$\xi(x,y)$		$\longrightarrow$

Table 5.4. Catalogue of First-Order ODEs

Item	Symbol	$X = u$	Y	dY/dX	Enlargement
Ia1	$M(x)f_x$	y	$\displaystyle\int \frac{dx}{M}$	$\dfrac{1}{My'}$	$My' + yq(Y) = g$
Ias1	$M(y)f_y$	x	$\displaystyle\int \frac{dy}{M}$	$\dfrac{y'}{M}$	$y' + MYq(x) = Ag$ $\dfrac{M}{y'} + xq(Y) = g$
Ib1	$N(y)f_x$	y	$\dfrac{x}{N}$	$\dfrac{N - xN'y'}{N^2 y'}$	$\dfrac{1}{y'} + xq(y) = g$ $\dfrac{N^2 y'}{N - xN'y'} + yq\left(\dfrac{x}{N}\right) = g$
Ibs1	$N(x)f_y$	x	$\dfrac{y}{N}$	$\dfrac{Ny' - N'y}{N^2}$	$y' + yq(x) = g$ $\dfrac{N^2}{Ny' - N'y} + xq\left(\dfrac{y}{N}\right) = g$
Ic1	$M(x)N(y)f_x$	y	$\dfrac{1}{N}\displaystyle\int \frac{dx}{M}$	$\dfrac{1 - MN'Yy'}{MNy'}$	$\dfrac{1}{y'} + MNYq(y) = Mg$ $\dfrac{MNy'}{1 - MN'Yy'} + yq(Y) = g$

Table 5.4. Catalogue of First-Order ODEs

Item	Symbol	$X=u$	Y	dY/dX	Enlargement
Ics1	$M(y)N(x)f_y$	x	$\dfrac{1}{N}\displaystyle\int\dfrac{dy}{M}$	$\dfrac{y'-MN'Y}{MN}$	$y' + M\left(\displaystyle\int\dfrac{dy}{M}\right)q(x) = Mg$ $\dfrac{MN}{y'-MN'Y}+xq(Y)=g$
Ic2	$x^a N(y)f_x$ $N=\exp\left[(a-1)\displaystyle\int P(y)dy\right]$ $a\neq 1$	y	$\dfrac{x^{1-a}}{N}$	$\dfrac{(1-a)(1+xPy')}{x^a N y'}$	$\dfrac{1}{y'}+xq(y)=x^a g$ $\dfrac{xy'}{1+xPy'}+yq\left(\dfrac{x^{1-a}}{N}\right)=g$
Ics2	$y^a N(x)f_y$ $N=\exp\left[(a-1)\displaystyle\int P(x)dx\right]$ $a\neq 1$	x	$\dfrac{y^{1-a}}{N}$	$\dfrac{(1-a)(y'+Py)}{N y^a}$	$y'+yq(x)=y^a g$ $\dfrac{y}{y'+Py}+xq\left(\dfrac{y^{1-a}}{N}\right)=g$
Ic3	$x^m y^n f_x$ $m\neq 1$	y	$\dfrac{x^{1-m}}{y^n}$	$\dfrac{(1-m)y-nxy'}{x^m y^{n+1}y'}$	$\dfrac{xyy'}{(1-m)y-nxy'}+yq\left(\dfrac{x^{1-m}}{y^n}\right)=g$

Table 5.4. Catalogue of First-Order ODEs

Item	Symbol	$X = u$	Y	dY/dX	Enlargement
Ics3	$x^n y^m f_y$ $m \neq 1$	x	$\dfrac{y^{1-m}}{x^n}$	$\dfrac{(1-m)xy' - ny}{x^{n+1}y^m}$	$\dfrac{xy}{\substack{(1-m)xy'^{1-m} - ny \\ +xq\left(\dfrac{y^{1-m}}{x^n}\right)}} = g$
Ic4	$xy^n f_x$	y	$\dfrac{\ln x}{y^n}$	$\dfrac{y - nx(\ln x)y'}{xy^{n+1}y'}$	$\dfrac{\frac{1}{y'} + x(\ln x)q(y) = xg}{\substack{y - nx(\ln x)y' \\ xy^{n+1}y'} \\ +yq\left(\dfrac{\ln x}{y^n}\right)} = g$
Ics4	$x^n y f_y$	x	$\dfrac{\ln y}{x^n}$	$\dfrac{xy' - ny\ln y}{x^{n+1}y}$	$\dfrac{y' + y(\ln y)q(x) = yg}{\substack{xy' - ny(\ln y) \\ x^{n+1}y} \\ +xq\left(\dfrac{\ln y}{x^n}\right)} = g$
Ic5	$\exp(mx+ny)f_x$	y	$\exp[-(mx+ny)]$	$-\dfrac{m+ny'}{y'}Y$	$\dfrac{\frac{1}{y'} + q(y) = e^{mx}g}{\dfrac{y'}{m+ny'} + yq(e^{mx+ny})} = g$

Table 5.4. Catalogue of First-Order ODEs

Item	Symbol	$X = u$	Y	dY/dX	Enlargement
Ics5	$\exp(nx+my)f_y$	x	$\exp[-(nx+my)]$	$-(n+my')Y$	$y' + q(x) = e^{my}g$ $\dfrac{1}{n+my'} + xq(e^{nx+my}) = g$
Ic6	$\exp(mx^2+ny^2)f_x$	y	$e^{-ny^2}erf(m^{1/2}x)$	$\dfrac{2e^{-ny^2}}{y'}\left[\left(\dfrac{m}{\pi}\right)^{1/2}e^{-mx^2} - nyy'erf(m^{1/2}x)\right]$	$\dfrac{\frac{1}{y'} + e^{mx^2}erf(m^{1/2}x)q(y) = e^{mx^2}g}{e^{ny^2}y'}$ $\left(\dfrac{m}{\pi}\right)^{1/2}\ e^{-mx^2} - nyy'erf(m^{1/2}x)$ $+yq[e^{-ny^2}erf(m^{1/2}x)] = g$
Ics6	$\exp(nx^2+my^2)f_y$	x	$e^{-nx^2}erf(m^{1/2}y)$	$2e^{-nx^2}\left[\left(\dfrac{m}{\pi}\right)^{1/2}e^{-my^2}y' - nxerf(m^{1/2}y)\right]$	$\dfrac{e^{-my^2}y' + erf(m^{1/2}y)q(x) = g}{e^{nx^2}}$ $\left(\dfrac{m}{\pi}\right)^{1/2}\ e^{-my^2}y' - nxerf(m^{1/2}y)$ $+xq[e^{-nx^2}erf(m^{1/2}y)] = g$

Table 5.4. Catalogue of First-Order ODEs

Item	Symbol	$X = u$	Y	dY/dX	Enlargement
Id1	$\xi(x,y)f_x$	x	$\displaystyle\int \frac{dx}{\xi(x,c)}$	$\displaystyle\frac{1}{\xi y'} - \int_{y=c} \frac{\partial \xi}{\partial y}\frac{dx}{\xi^2}$	$\displaystyle\frac{1}{\xi y'} - \int_{y=c} \frac{\partial \xi}{\partial y}\frac{dx}{\xi^2} + Y q(x) = g$
Ids1	$\xi(y,x)f_y$	y	$\displaystyle\int \frac{dy}{\xi(c,y)}$	$\displaystyle\frac{y'}{\xi} - \int_{x=c} \frac{\partial \xi}{\partial x}\frac{dy}{\xi^2}$	$\displaystyle\frac{y'}{\xi} - \int_{x=c} \frac{\partial \xi}{\partial x}\frac{dy}{\xi^2} + y q(Y) = g$
Id2	$A(xy)f_x$	y	$\displaystyle\int \frac{dx}{A(cx)}$	$\displaystyle\frac{1}{Ay'} - \int \frac{\partial A}{\partial y}\frac{dx}{A^2}$	$\displaystyle\frac{1}{Ay'} - \int \frac{\partial A}{\partial y}\frac{dx}{A^2} + Y q(y) = g$
Ids2	$A(xy)f_y$	x	$\displaystyle\int \frac{dy}{A(cy)}$	$\displaystyle\frac{y'}{A} - \int \frac{\partial A}{\partial x}\frac{dy}{A^2}$	$\displaystyle\frac{y'}{A} - \int \frac{\partial A}{\partial x}\frac{dy}{A^2} + Y q(x) = g$
Id3	$x(x+y)f_x$	y	$\displaystyle\frac{1}{y}\ln\left(\frac{x}{x+y}\right)$	$\displaystyle-\frac{1}{y^2}\ln\left(\frac{x}{x+y}\right) - \frac{xy'-y}{xy(x+y)y'}$	$\displaystyle\frac{xy'-y}{x(x+y)y'} + \ln\left(\frac{x+y}{x}\right) q(y) = g$

$$\frac{1}{\dfrac{xy'-y}{x(x+y)y'} + \dfrac{1}{y}\ln\left(\dfrac{x}{x+y}\right)}$$

$$+ q\left(\frac{1}{y}\ln\frac{x+y}{x}\right) = \frac{g}{y}$$

Table 5.4. Catalogue of First-Order ODEs

Item	Symbol	$x = u$	Y	dY/dX	Enlargement
Ids3	$y(x+y)f_y$	x	$\dfrac{1}{x}\ln\left(\dfrac{y}{x+y}\right)$	$-\dfrac{1}{x^2}\ln\left(\dfrac{y}{x+y}\right) - \dfrac{y-xy'}{xy(x+y)}$	$\dfrac{y-xy'}{(x+y)y} + \ln\left(\dfrac{x+y}{y}\right)q(x) = g$
					$\dfrac{1}{\dfrac{y-xy'}{(x+y)y} + \dfrac{1}{x}\ln\left(\dfrac{y}{x+y}\right)}$
					$+q\left(\dfrac{1}{x}\ln\dfrac{x+y}{y}\right) = \dfrac{g}{x}$
Id4	$x(x-y)f_x$	y	$\dfrac{1}{y}\ln\dfrac{x-y}{x}$	$\dfrac{y-xy'}{(x-y)xyy'} + \dfrac{1}{y^2}\ln\dfrac{x}{x-y}$	$\dfrac{y-xy'}{x(x-y)y'} + \ln\left(\dfrac{x}{x-y}\right)q(y) = g$
					$\dfrac{1}{\dfrac{y-xy'}{x(x-y)y'} + \dfrac{1}{y}\ln\left(\dfrac{x}{x-y}\right)}$
					$+q\left[\dfrac{1}{y}\ln\left(\dfrac{y-x}{x}\right)\right] = \dfrac{g}{y}$

Table 5.4. Catalogue of First-Order ODEs

Item	Symbol	$x = u$	Y	dY/dX	Enlargement
Ids4	$y(y-x)f_y$	x	$\dfrac{1}{x}\ln\dfrac{y-x}{y}$	$\dfrac{xy'-y}{xy(y-x)}+\dfrac{1}{x^2}\ln\dfrac{y}{y-x}$	$\dfrac{xy'-y}{y(y-x)}+\ln\left(\dfrac{y}{y-x}\right)q(x)=g$ $$\dfrac{1}{\dfrac{xy'-y}{y(y-x)}+\dfrac{1}{x}\ln\left(\dfrac{y}{y-x}\right)}$$ $$+q\left[\dfrac{1}{x}\ln\left(\dfrac{y-x}{y}\right)\right]=\dfrac{g}{x}$$
Id5	$e^{axy}f_x$	y	$\dfrac{1}{y}e^{-axy}$	$-\dfrac{ay^2+(1+axy)y'}{y^2y'}e^{-axy}$	$\dfrac{y+axy'}{y'}+q(y)=e^{axy}g$ $$\dfrac{1}{ay^2+(1+axy)y'}+q(ye^{axy})=\dfrac{g}{y}$$

Table 5.4. Catalogue of First-Order ODEs

Item	Symbol	$X = u$	Y	dY/dX	Enlargement
Ids5	$e^{axy}f_y$	x	$\dfrac{1}{x}e^{-axy}$	$-\dfrac{1+axy+ax^2y'}{x^2}e^{-axy}$	$xy'+y+q(x)=e^{axy}g$
					$\dfrac{1}{1+axy+ax^2y'}+q(xe^{axy})=\dfrac{g}{x}$
Id6	$\dfrac{1}{xy}e^{axy}f_x$	y	$\dfrac{1+axy}{y}e^{-axy}$	$-\dfrac{a^2xy^3+(1+axy+a^2x^2y^2)y'}{e^{axy}y^2y'}$	$\dfrac{x(y+xy')}{y'}+(1+axy)q(y)=e^{axy}g$
					$\dfrac{(1+axy)y'}{a^2xy^3+(1+axy+a^2x^2y^2)}$
					$+q\left(\dfrac{ye^{axy}}{1+axy}\right)=\dfrac{g}{y}$
Ids6	$\dfrac{1}{xy}e^{axy}f_y$	x	$\dfrac{1+axy}{x}e^{-axy}$	$-\dfrac{a^2x^3yy'+(1+axy+a^2x^2y^2)}{e^{axy}x^2}$	$(xy'+y)y+(1+axy)q(x)=e^{axy}g$
					$\dfrac{1+axy}{a^2x^3yy'+(1+axy+a^2x^2y^2)}$
					$+q\left(\dfrac{xe^{axy}}{1+axy}\right)=g$

Table 5.4. Catalogue of First-Order ODEs

Item	Symbol	$X = u$	Y	dY/dX	Enlargement
IIa1	$M(x)f_x + A(x)f_y$	$\int \frac{A}{M}dx - y$	$\int \frac{dx}{M}$	$\frac{1}{A - My'}$	$\frac{1}{My' - A} + Yq(X) = g$
					$My' + Xq(Y) = g$
IIas1	$A(y)f_x + M(y)f_y$	$-x + \int \frac{A}{M}dy$	$\int \frac{dy}{M}$	$\frac{y'}{Ay' - M}$	$\frac{y'}{Ay' - M} + Yq(X) = g$
					$\frac{M}{y'} + Xq(Y) = g$
IIa2	$af_x + bf_y$	$\frac{b}{a}x - y$	x	$\frac{a}{b - ay'}$	$\frac{1}{ay' - b} + xq(bx - ay) = g$
IIa3	$ay^m f_x + byf_y$ $m \neq 0$	$\frac{b}{a}x - \frac{y^m}{m}$	$\ln\left(\frac{a}{b}\frac{y^m}{m}\right)$	$\frac{amy'}{by - ay^my'}$	$\frac{y'}{ay^my' - by} + Yg(X) = g$
IIb1	$N(y)f_x + A(x)f_y$	$-\int Adx + \int Ndy$	$\int \frac{dx}{N}$	$\frac{\frac{1}{N} - y'\int \frac{N'dx}{N^2}}{Ny' - A}$	$\frac{\frac{1}{N} - y'\int \frac{N'dx}{N^2}}{Ny' - A} + Yq(X) = g$

Table 5.4. Catalogue of First-Order ODEs

Item	Symbol	$X = u$	Y	dY/dX	Enlargement
IIbs1	$A(y)f_x$ $+N(x)f_y$	$\int N\,dx$ $-\int A\,dy$	$\int \dfrac{dx}{A}$	$\dfrac{1}{A(N-Ay')}$	$\dfrac{1}{A(N-Ay')}+Yq(X)=g$ $A(N-Ay')+Xq(Y)=g$
IIb2	$ayf_x + bxf_y$ $ab>0$	$bx^2 - ay^2$	$\ln\left[y + \left(\dfrac{b}{a}\right)^{1/2} x\right]$	$-\dfrac{(ab)^{1/2}(xy'-y)-(ayy'-bx)}{2(bx^2-ay^2)(ayy'-bx)}$	$\dfrac{xy'-y}{ayy'-bx}+Yq(\cdot X)=g$ $\dfrac{ayy'-bx}{(ab)^{1/2}(xy'-y)-(ayy'-bx)}$ $+q(b^{1/2}x+a^{1/2}y)$ $=\dfrac{g}{bx^2-ay^2}$

Table 5.4. Catalogue of First-Order ODEs

Item	Symbol	$X = u$	Y	dY/dX	Enlargement
IIb3	$ayf_x - bxf_y$ $ab > 0$	$bx^2 + ay^2$	$\tan^{-1}$ $\times\left[\left(\dfrac{a}{b}\right)^{1/2}\dfrac{y}{x}\right]$	$\dfrac{(ab)^{1/2}(xy'-y)}{2(bx^2+ay^2)(ayy'+bx)}$	$\dfrac{xy'-y}{ayy'+bx} + Yg(X) = g$ $\dfrac{ayy'+bx}{xy'-y} + q\left(\dfrac{y}{x}\right)$ $= \dfrac{g}{bx^2+ay^2}$
IIc1	$M(x)N(y)f_x$ $+A(x)f_y$	$\displaystyle\int \dfrac{A}{M}dx$ $-\displaystyle\int Ndy$	$\displaystyle\int \dfrac{dy}{A(x)}$	$\dfrac{y' - \displaystyle\int \dfrac{A'}{A^2}dy}{\dfrac{A}{M} - Ny'}$	$\dfrac{y' - \displaystyle\int \dfrac{A'}{A^2}dy}{\dfrac{A}{M} - Ny'} + Yq(X) = g$

Table 5.4. Catalogue of First-Order ODEs

Item	Symbol	$X = u$	Y	dY/dX	Enlargement
IIcs1	$A(y)f_x$ $+M(y)N(x)f_y$	$\int N dx$ $-\int \dfrac{A}{M} dy$	$\int \dfrac{dx}{A(y)}$	$\dfrac{\frac{1}{A} - y' \int \frac{A'}{A^2} dx}{N - \frac{A}{M} y'}$	$\dfrac{\frac{1}{A} - y' \int \frac{A'}{A^2} dx}{N - \frac{A}{M} y'}$ $+Y q(X) = g$
IIc2	$-\dfrac{1+x}{y} f_x + a f_y$	$(1+x)^a y$	y	$\dfrac{(1+x)^{1-a} y'}{(1+x)y' + ay}$	$\dfrac{(1+x)y'}{(1+x)y' + ay}$ $+yg[(1+x)^a y] = g$ $\dfrac{1}{(1+x)y'} + q(y) = \dfrac{g}{(1+x)^a}$
IIcs2	$af_x - \dfrac{1+y}{x} f_y$	$x(1+y)^a$	x	$\dfrac{(1+y)^{1-a}}{1+y+axy'}$	$\dfrac{1+y}{1+y+axy'} + q[x(1+y)^a] = \dfrac{g}{x}$ $\dfrac{y'}{1+y'} + q(x) = \dfrac{g}{(1+y)^a}$

Table 5.4. Catalogue of First-Order ODEs

Item	Symbol	$X = u$	Y	dY/dx	Enlargement
IId1	$(x+y)f_x + af_y$	$(x+y+a)e^{-y/a}$	$\dfrac{y}{a}$	$\dfrac{e^{y/a}y'}{a-(x+y)y'}$	$\dfrac{e^{y/a}y'}{a-(x+y)y'} + yq(X) = g$
					$\dfrac{a-(x+y)y'}{y'}$ $+(x+y+a)q(y) = g$
IIds1	$af_x + (x+y)f_y$	$(x+y+a)e^{-x/a}$	$\dfrac{x}{a}$	$\dfrac{e^{x/a}}{ay'-(x+y)}$	$\dfrac{e^{x/a}}{ay'-(x+y)} + xq[(x+y+a)e^{-x/a}] = g$
					$ay' + (x+y+a)q(x) = g+y$
IId2	$\dfrac{1}{x+y}f_x + \dfrac{1}{x}f_y$	$\dfrac{y}{x}-\ln x$	$\dfrac{1}{4}x^2 + \dfrac{1}{2}xy$	$\dfrac{x^2}{2}\dfrac{xy'+(x+y)}{xy'-(x+y)}$	$\dfrac{xy'+(x+y)}{xy'-(x+y)}$
					$+\dfrac{x+2y}{x}q\left(\dfrac{y}{x}-\ln x\right) = \dfrac{1}{x^2}g$
					$\dfrac{xy'-(x+y)}{xy'+(x+y)} + x^2\,Xq(Y) = x^2 g$

Table 5.4. Catalogue of First-Order ODEs

Item	Symbol	$X = u$	Y	dY/dx	Enlargement
IIds2	$\dfrac{1}{y}f_x + \dfrac{1}{x+y}f_y$	$\dfrac{x}{y} - \ln y$	$\dfrac{1}{4}y^2 + \dfrac{1}{2}xy$	$\dfrac{y^2}{2}\dfrac{y+(x+y)y'}{y-(x+y)y'}$	$\dfrac{y+(x+y)y'}{y-(x+y)y'}$

$$+\frac{y+2x}{y}q\left(\frac{x}{y}-\ln y\right) = \frac{1}{y^2}g$$

$$\frac{y-(x+y)y'}{y+(x+y)y'}$$

$$+y^2\left(\frac{x}{y}-\ln y\right)q(y^2+2xy) = y^2g$$

Table 5.4. Catalogue of First-Order ODEs

Item	Symbol	$X = u$	Y	dY/dx	Enlargement
IIIa1	$M(x)f_x + B(y)f_y$	$\int \dfrac{dx}{M} - \int \dfrac{dy}{B}$	$\int \dfrac{dx}{M}$	$\dfrac{B}{B-My'}$	$\dfrac{B}{B-My'} + Yq(X) = g$
					$\dfrac{M}{B}y' + Xq(Y) = g$
IIIa2	$axf_x + by^r f_y,\ r \neq 1$	$\dfrac{b}{a}\ln x - \dfrac{y^{1-r}}{1-r}$	$\ln x$	$\dfrac{ay^r}{by^r - axy'}$	$\dfrac{y^r}{axy' - by^r} + Yq(X) = g$
IIIas2	$bx^r f_x + ayf_y,\ r \neq 1$	$\dfrac{b}{a}\ln y - \dfrac{x^{1-r}}{1-r}$	$\ln y$	$\dfrac{ax^r y'}{bx^r y' - ay}$	$\dfrac{x^r y'}{bx^r y' - ay} + Yq(X) = g$
IIIa3	$axf_x + byf_y$	$\dfrac{x^b}{y^a}$	$\ln x$	$\dfrac{y^{a+1}}{x^b(by - axy')}$	$\dfrac{y}{axy' - by} + (\ln x)q\left(\dfrac{x^b}{y^a}\right) = g$
IIIc1	$M(x)N(y)f_x + B(y)f_y$	$\int \dfrac{N}{B}dy - \int \dfrac{dx}{M}$	$\int \dfrac{dy}{B}$	$\dfrac{My'}{NMy' - B}$	$\dfrac{My'}{MNy' - B} + Yq(X) = g$
					$\dfrac{MNy' - B}{My'} + Xq(Y) = g$

Table 5.4. Catalogue of First-Order ODEs

Item	Symbol	$X = u$	Y	dY/dX	Enlargement
IIIcs1	$B(x)f_x + M(y)N(x)f_y$	$\int \frac{N}{B}dx - \int \frac{dy}{M}$	$\int \frac{dy}{B}$	$\frac{M}{MN - By'}$	$\frac{M}{MN - By'}$ $+Yq(X) = g$ $y' + MXq(Y') = Mg$
IIIc2	$bx^r y^n f_x + ay^m f_y$ $n \neq m-1,\ m \neq 1,\ r \neq 1$	$\frac{by^{n-m+1}}{a(n-m+1)} - \frac{x^{1-r}}{1-r}$	y^{1-m}	$\frac{a(m-1)x^r y'}{ay^m - bx^r y^n y'}$	$\frac{x^r y'}{bx^r y^n - my' - a}$ $+yq(X) = y^m g$
IIIcs2	$ax^m f_x + bx^n y^r f_y$ $n \neq m-1,\ m \neq 1,\ r \neq 1$	$\frac{bx^{n-m+1}}{a(n-m+1)} - \frac{y^{1-r}}{1-r}$	x^{1-m}	$\frac{a(m-1)y^r}{ax^m y' - bx^n y^r}$	$\frac{y^r}{ay' - bx^{n-m}y^r}$ $+xq(X) = x^m g$
IIIc3	$bx^r y^n f_x + ay^{n+1} f_y$ $n \neq 0,\ r \neq 1$	$\frac{b}{a}\ln y - \frac{x^{1-r}}{1-r}$	y^{-n}	$\frac{anx^r y'}{ay^{n+1} - bx^r y^n y'}$	$\frac{x^r y'}{bx^r y' - ay}$ $+q(X) = g$
IIIcs3	$ax^{n+1} f_x + bx^n y^r f_y$ $n \neq 0,\ r \neq 1$	$\frac{b}{a}\ln x - \frac{y^{1-r}}{1-r}$	x^{-n}	$\frac{any^r}{ax^{n+1}y' - bx^n y^r}$	$\frac{y^r}{axy' - by^r}$ $+q(X) = g$

Table 5.4. Catalogue of First-Order ODEs

Item	Symbol	$X = u$	Y	dY/dX	Enlargement
IIIc4	$bxy^n f_x + ay^{n+1}f_y$ $n \neq 0$	$\dfrac{y^b}{x^a}$	y^{-n}	$\dfrac{nx^{a+1}y'}{y^{b+n}(ay - bxy')}$	$\dfrac{xy'}{bxy' - ay} + q\left(\dfrac{y^b}{x^a}\right) = y^n g$
IIIcs4	$ax^{n+1}f_x + bx^n y f_y$ $n \neq 0$	$\dfrac{x^b}{y^a}$	x^{-n}	$\dfrac{ny^{a+1}}{x^{b+n}(axy' - by)}$	$\dfrac{y}{axy' - by} + q\left(\dfrac{x^b}{y^a}\right) = x^n g$
IIIc5	$bx^r y^n f_x + ayf_y$ $n \neq 0,\ r \neq 1$	$\dfrac{b}{a}\dfrac{y^n}{n} - \dfrac{x^{1-r}}{1-r}$	$\ln y$	$\dfrac{ax^r y'}{bx^r y^n y' - ay}$	$\dfrac{x^r y'}{bx^r y^n y' - ay} + (\ln y)q(X) = g$
IIIcs5	$axf_x + bx^n y^r f_y$ $n \neq 0,\ r \neq 1$	$\dfrac{b}{a}\dfrac{x^n}{n} - \dfrac{y^{1-r}}{1-r}$	$\ln x$	$\dfrac{ay^r}{bx^n y^r - axy'}$	$\dfrac{y^r}{axy' - bx^n y^r} + (\ln x)q(X) = g$
IIIc6	$bxy^n f_x + ayf_y$ $n \neq 0$	$\dfrac{b}{a}\dfrac{y^n}{n} - \ln x$	$\ln y$	$\dfrac{axy'}{bxy^n y' - ay}$	$\dfrac{xy'}{bxy^n y' - ay}$ $+ (\ln y)q\left(\dfrac{b}{a}\dfrac{y^n}{n} - \ln x\right) = g$

Table 5.4. Catalogue of First-Order ODEs

Item	Symbol	$X = u$	Y	dY/dX	Enlargement
IIIcs6	$axf_x + bx^n y f_y$ $n \neq 0$	$\dfrac{b}{a}\dfrac{x^n}{n} - \ln y$	$\ln x$	$\dfrac{ay}{bx^n y - axy'}$	$\dfrac{y}{axy' - bx^n y}$ $+ (\ln x) q\left(\dfrac{b}{a}\dfrac{x^n}{n} - \ln y\right) = g$
IIIc7	$bxy^n f_x + ay^m f_y$ $n \neq m-1,\ m \neq 1$	$\dfrac{b}{a}\dfrac{y^{n-m+1}}{n-m+1} - \ln x$	y^{1-m}	$\dfrac{a(1-m)xy'}{bxy^n y' - ay^m}$	$\dfrac{xy'}{bxy^{n-m}y' - a}$ $+ yq(X) = y^m g$
IIIcs7	$ax^m f_x + bx^n y f_y$ $n \neq m-1,\ m \neq 1$	$\dfrac{bx^{n-m-1}}{a(n-m+1)} - \ln y$	x^{1-m}	$\dfrac{a(m-1)y}{ax^m y' - bx^n y}$	$\dfrac{y}{ay' - bx^{n-m}y}$ $+ xq(X) = x^m g$
IIIc8	$\dfrac{y^n}{P(y)}(axf_x + yf_y)$	$\dfrac{y^a}{x}$	$\displaystyle\int \dfrac{Pdy}{y^{n+1}}$	$\dfrac{x^2 Py'}{y^{n+a}(axy' - y)}$	$\dfrac{xPy'}{y^n(axy' - y)} + Yq\left(\dfrac{y^a}{x}\right) = g$
IIIcs8	$\dfrac{x^n}{P(x)}(xf_x + ayf_y)$	$\dfrac{y}{x^a}$	$\displaystyle\int \dfrac{Pdx}{x^{n+1}}$	$\dfrac{x^{a-n}P}{xy' - ay}$	$\dfrac{x^{a-n}P}{xy' - ay} + Yq\left(\dfrac{y}{x^a}\right) = g$

Table 5.4. Catalogue of First-Order ODEs

Item	Symbol	$X = u$	Y	dY/dX	Enlargement
IIId1	$(x+y)f_x + yf_y$	$\dfrac{x}{y} - \ln y$	$\ln y$	$\dfrac{yy'}{y - (x+y)y'}$	$\dfrac{yy'}{y - (x+y)y'} + (\ln y)q\left(\dfrac{x}{y} - \ln y\right) = g$
					$\dfrac{y - (x+y)y'}{y'} + \left(\dfrac{x}{y} - \ln y\right)q(y) = g$
IIIds1	$xf_x + (x+y)f_y$	$\dfrac{y}{x} - \ln x$	$\ln x$	$\dfrac{x}{xy' - (x+y)}$	$\dfrac{x}{xy' - (x+y)} + (\ln x)q\left(\dfrac{y}{x} - \ln x\right) = g$
					$xy' + \left(\dfrac{y}{x} - \ln x\right)q(x) = g + y$

Table 5.4. Catalogue of First-Order ODEs

Item	Symbol	$X = u$	Y	dY/dX	Enlargement
IVc1	$x^m y^n (axf_x + byf_y)$ $am + bn \neq 0$	$\dfrac{y^a}{x^b}$	$\dfrac{1}{x^m y^n}$	$-\left(\dfrac{x^b}{y^a}\right) \dfrac{1}{x^m y^n} \dfrac{nxy' + my}{axy' - by}$	$\dfrac{nxy' + my}{axy' - by} + q\left(\dfrac{y^a}{x^b}\right) = x^m y^n g$ $\dfrac{axy' - by}{nxy' + my} + q(x^m y^n) = \dfrac{x^b}{y^a} g$
IVc2	$\left(\dfrac{y}{x^{b/a}}\right)^n (axf_x + byf_y)$	$\dfrac{y^a}{x^b}$	$\left(\dfrac{x^{b/a}}{y}\right)^n \ln y$	$\left(\dfrac{x^{b/a}}{y}\right)^n \dfrac{x^b}{y^a}$ $\times \dfrac{(1 - n\ln y)xy' + \dfrac{bn}{a} y \ln y}{axy' - by}$	$\dfrac{(1 - n\ln y)xy' + \dfrac{bn}{a} y \ln y}{axy' - by}$ $+ (\ln y) q\left(\dfrac{y^a}{x^b}\right) = g$ $\dfrac{(axy' - by)\ln y}{(1 - n\ln y)xy' + \dfrac{bn}{a} y \ln y}$ $+ q\left[\left(\dfrac{x^{b/a}}{y}\right)\ln y\right] = \dfrac{x^b}{y^a} g$

Table 5.4. Catalogue of First-Order ODEs

Item	Symbol	$X = u$	Y	dY/dX	Enlargement
IVc3	$xy(af_x + bf_y)$	$bx - ay$	$\dfrac{1}{bx-ay}\ln\dfrac{ay}{bx}$	$\dfrac{1}{xy(bx-ay)}\dfrac{xy'-y}{b-ay'} - \dfrac{\ln(ay/bx)}{(bx-ay)^2}$	$\dfrac{1}{xy}\dfrac{xy'-y}{b-ay'} + \ln\left(\dfrac{ay}{bx}\right)q(bx-ay) = g$
					$\dfrac{\ln(ay/bx)}{\left(\dfrac{bx-ay}{b-ay'}\right)\dfrac{xy'-y}{xy} - \ln(ay/bx)} + q(Y) = g$
IVc4	$\dfrac{1}{xy}(af_x + bf_y)$	$bx - ay$	$x^2(bx - 3ay)$	$\dfrac{3x^2(b-ay')-6axy}{b-ay'}$	$\dfrac{2ay}{b-ay'}$
					$+x(bx-3ay)q(bx-ay) = \dfrac{g}{x} + x$
					$\dfrac{b-ay'}{x(b-ay')-2ay}$
					$+x(bx-ay)q(bx^3 - 3ax^2y) = xg$

Table 5.4. Catalogue of First-Order ODEs

Item	Symbol	$X=u$	Y	dY/dX	Enlargement
IVd1	$\dfrac{1}{x+y}f_x + \dfrac{y}{x^2}f_y$	$-\dfrac{x}{y}-\ln x$	$\displaystyle\int (x+y)dx$ $= \dfrac{1}{2}x^2 - \displaystyle\int \dfrac{x\,dx}{c+\ln x}$	$\dfrac{x(x+y)y^2}{x^2y' - (x+y)y}$	$\dfrac{x(x+y)y^2}{x^2y'-(x+y)y}$ $+Yq\left(\dfrac{x}{y}+\ln x\right)=g$ $\dfrac{x^2y'-(x+y)y}{x(x+y)y^2}$ $+\left(\dfrac{x}{y}+\ln x\right)q(Y)=g$
IVds1	$\dfrac{x}{y^2}f_x + \dfrac{1}{x+y}f_y$	$-\dfrac{y}{x}-\ln y$	$\displaystyle\int (x+y)dy$ $= \dfrac{1}{2}y^2 - \displaystyle\int \dfrac{y\,dy}{c+\ln y}$	$\dfrac{x^2(x+y)yy'}{y^2 - x(x+y)y'}$	$\dfrac{x^2(x+y)yy'}{y^2-x(x+y)y'}$ $+Yq\left(\dfrac{y}{x}+\ln y\right)=g$ $\dfrac{y^2-x(x+y)y'}{x^2(x+y)yy'}$ $+\left(\dfrac{y}{x}+\ln y\right)q(Y)=g$

Table 5.4. Catalogue of First-Order ODEs

Item	Symbol	$X = u$	Y	dY/dX	Enlargement
V1	$\dfrac{P_y f_x - P_x f_y}{P_y Q_x - P_x Q_y}\,e(Q)$	$P(x,y)$	$\displaystyle\int \dfrac{dQ}{E(Q)}$	$\dfrac{1}{E}\dfrac{Q_x + Q_y y'}{P_x + P_y y'}$	$\dfrac{1}{E}\dfrac{Q_x + Q_y y'}{P_x + P_y y'}$
					$+Yq(P) = g$
					$E\dfrac{P_x + P_y y'}{Q_x + Q_y y'}$
					$+Pq(Y) = g$
V2	$P = x^a y^b,\ Q = x^m y^n,\quad x^a y^b$	$\ln(x^m y^n)$		$\dfrac{1}{x^a y^b}\dfrac{nxy' + my}{bxy' + ay}$	$\dfrac{nxy' + my}{bxy' + ay}$
	$E = Q,\ an \neq bm$				$+\ln(x^m y^n)q(x^a y^b) = g$
					$\dfrac{bxy' + ay}{nxy' + my}$
					$+q(x^m y^n) = \dfrac{g}{x^a y^b}$

180

Table 5.4. Catalogue of First-Order ODEs

Item	Symbol	$X = u$	Y	dY/dX	Enlargement
V3	$P = x^a y^b$,	$x^a y^b$	$-\exp(-x^m y^n)$	$\dfrac{x^{m-a} y^{n-b}}{\exp(x^m y^n)} \dfrac{nxy' + my}{bxy' + ay}$	$x^m y^n \dfrac{nxy' + my}{bxy' + ay}$
	$Q = x^m y^n$,				$+ q(x^a y^b) = \exp(x^m y^n)g$
	$E = \exp(x^m y^n)$,				
	$an \neq bm$				
V4	$x(x+y)f_x$	xy	$\dfrac{1}{(xy)^{1/2}}$ $\times \tan^{-1}\left(\dfrac{x}{y}\right)^{1/2}$	$\dfrac{(xy)^{1/2} - (x+y)\tan^{-1}\left(\dfrac{x}{y}\right)^{1/2}}{2(x+y)(xy)^{3/2}}$ $\times \dfrac{y - xy'}{y + xy'}$	$\dfrac{(xy)^{1/2} - (x+y)\tan^{-1}\left(\dfrac{x}{y}\right)^{1/2}}{x + y}$ $\times \dfrac{y - xy'}{y + xy'} + \tan^{-1}\left(\dfrac{x}{y}\right)^{1/2} q(X) = g$
	$-y(x+y)f_y$				$\dfrac{(xy)^{1/2} - (x+y)\tan^{-1}\left(\dfrac{x}{y}\right)^{1/2}}{(x + y)(xy)^{3/2}}$ $\times \dfrac{y + xy'}{y - xy'} + xyq(Y) = g$

Table 5.4. Catalogue of First-Order ODEs

Item	Symbol	$X = u$	Y	dY/dX	Enlargement
Vs4	$-x(x+y)f_x$ $+y(x+y)f_y$	$\dfrac{y}{x}$	$\dfrac{1}{x+y}$	$\left(\dfrac{x}{x+y}\right)^2 \dfrac{1+y'}{y-xy'}$	$\dfrac{1+y'}{\frac{y}{x}-y'} + q\left(\dfrac{y}{x}\right) = xg$ $\dfrac{y-xy'}{1+y'}$ $+xyq(x+y) = x^2g$
V5	$x(x+y)f_x$ $+y(x+y)f_y$	$\dfrac{y}{x}$	$\dfrac{1}{x+y}$	$-\left(\dfrac{x}{x+y}\right)^2 \dfrac{1+y'}{xy'-y}$	$\dfrac{1+y'}{xy'-y} + \dfrac{1}{x}q\left(\dfrac{y}{x}\right) = g$ $\dfrac{xy'-y}{1+y'} + xyq(x+y) = g$
V6	$x(x-y)f_x$ $+y(x-y)f_y$	$\dfrac{y}{x}$	$\dfrac{1}{x-y}$	$-\left(\dfrac{x}{x-y}\right)^2 \dfrac{1-y'}{xy'-y}$	$\dfrac{1-y'}{xy'-y} + \dfrac{1}{x-y}q\left(\dfrac{y}{x}\right) = g$ $\dfrac{xy'-y}{1-y'} + xyq(x-y) = x^2g$

Table 6.1. Catalogue of Second-Order ODEs

Item	d^2Y/dX^2	Simplified ODE
Ia1	$-\dfrac{My'' + M'y'}{M^2(y')^3}$	$M^2y'' + MM'y' = g(y, My')$
Ias1	$\dfrac{My'' - M'(y')^2}{M^2}$	$My'' - M'(y')^2 = M^2 g\left(x, \dfrac{y'}{M}\right)$
Ib1	$-\dfrac{1}{N}\left[\dfrac{y''}{(y')^3} + x\dfrac{N''}{N} + 2\dfrac{N'}{N}\left(\dfrac{N - xN'y'}{Ny'}\right)\right]$	$\dfrac{y''}{(y')^3} + x\dfrac{N''}{N} = g\left(y, \dfrac{N - xN'y'}{y'}\right)$
Ibs1	$\dfrac{1}{N^3}[N^2y'' - 2NN'y' + (2N'^2 - NN'')y]$	$Ny'' - N''y = g\left(x, y' - \dfrac{N'}{N}y\right)$
Ic1	$-\dfrac{1}{M^2N^2(y')^3}[MNy'' + M^2Y(NN'' - 2N'^2)(y')^3$ $+\, 2MN'(y')^2 + M'Ny']$	$y'' + MY\left(N'' - 2\dfrac{N'^2}{N}\right)(y')^3 + 2\dfrac{N'}{N}(y')^2 + \dfrac{M'}{M}y'$ $= M(y')^3 g\left(y, \dfrac{1 - MN'Yy'}{My'}\right)$
Ics1	$-\dfrac{1}{M^2N^2}[-MNy'' + M^2(NN'' - 2N'^2)Y$ $+\, 2MN'y' + M'N(y')^2]$	$y'' - MY\left(N'' - 2\dfrac{N'^2}{N}\right) - 2\dfrac{N'}{N}y' - \dfrac{M'}{M}(y')^2$ $= Mg\left(x, \dfrac{y' - MN'Y}{M}\right)$

Table 6.1. Catalogue of Second-Order ODEs

Item	d^2Y/dX^2	Simplified ODE
Ic2	$\dfrac{a-1}{x^a N (y')^2}\left[y'' + 2(a-1)y' + \dfrac{a}{x}\right]$	$y'' + 2(a-1)(y')^2 + \dfrac{a}{x}y'$ $= x^a(y')^3 g\left(y,\ \dfrac{1+xPy'}{x^a y'}\right)$
Ics2	$\dfrac{1-a}{N y^a}\left[y'' + 2(1-a)y' - \dfrac{a}{y}(y')^2\right]$	$y'' + 2(1-a)y' - \dfrac{a}{y}(y')^2$ $= y^a g\left(x,\ \dfrac{y'+yP}{y^a}\right)$
Ic3	$\dfrac{(m-1)xy^2 y'' + n(n+1)x^2(y')^3 + 2n(m-1)xy(y')^2 + m(m-1)y^2 y'}{x^{m+1}y^{n+2}(y')^3}$	$y^2 y'' + \dfrac{n(n+1)}{m-1}x(y')^3 + 2ny(y')^2 + \dfrac{my^2}{x}y'$ $= x^m(y')^3 g\left(y,\ \dfrac{(m-1)y+nxy}{x^m y'}\right)$
Ics3	$\dfrac{-(m-1)x^2 yy'' + n(n+1)y^2 + 2n(m-1)xyy' + m(m-1)x^2(y')^2}{x^{n+2}y^{m+1}}$	$-x^2 y'' + \dfrac{mx^2}{y}(y')^2 + 2nxy' + \dfrac{n(n+1)}{m-1}y$ $= y^m g\left(x,\ \dfrac{(m-1)xy'+ny}{y^m}\right)$

Table 6.1. Catalogue of Second-Order ODEs

Item	d^2Y/dX^2	Simplified ODE
Ic4	$-\dfrac{1}{x^2 y^n (y')^3}\left[xy'' - n(n+1)\dfrac{x^2 \ln x}{y^2}(y')^3\right]$ $+\dfrac{2nx}{y}(y')^2 + y'$	$xy^2 y'' - n(n+1)x^2(\ln x)(y')^3 + 2nxy(y')^2 + y^2 y'$ $= x^2(y')^3\, g\left(y,\ \dfrac{y - nx(\ln x)y'}{xy'}\right)$
Ics4	$\dfrac{1}{x^n y^2}\left[yy'' + n(n+1)\dfrac{y^2 \ln y}{x^2} - \dfrac{2ny}{x}y' - (y')^2\right]$	$x^2 yy'' + n(n+1)y^2 \ln y - 2nxyy' - x^2(y')^2$ $= y^2\, g\left(x,\ \dfrac{xy' - ny\ln y}{y}\right)$
Ic5	$\dfrac{Y}{(y')^3}[my'' + (m+ny')^2 y']$	$my'' + (m+ny')^2 y' = \dfrac{(y')^3}{Y}\, g(y, Y')$
Ics5	$Y[-my'' + (n+my')^2]$	$my'' - (n+my')^2 = g(x, Y')$
Id5	$\dfrac{e^{-axy}}{y^3(y')^3}\{[(1+axy)^2+1](y')^3 + 2a^2xy^3(y')^2$ $+a^2y^4y' + y(y' + ay^2)y''\}$	$y(y'+ay^2)y'' + [(1+axy)^2+1](y')^3 + 2a^2xy^3(y')^2$ $+a^2y^4y' = y^3(y')^3 e^{axy} g(y, Y')$

Table 6.1. Catalogue of Second-Order ODEs

Item	d^2Y/dX^2	Simplified ODE
IIa1	$\dfrac{M(My'' + M'y' - A')}{(A - My')^3}$	$My'' + M'y' - A' = \dfrac{1}{M}g(X, My' - A)$
IIas1	$-\dfrac{M}{(Ay' - M)^3}[My'' + (A'y' - M')(y')^2]$	$My'' + (A'y' - M')(y')^2 = \dfrac{(y')^3}{M}g\left(X, \dfrac{y'}{Ay' - M}\right)$
IIa2	$\dfrac{a^3 y''}{(b - ay')^3}$	$y'' = g(bx - ay', y')$
IIa3	$\dfrac{a^2 m[by^2 y'' + amy^m(y')^3 - by(y')^3]}{(by - ay^m y')^3}$	$by^2 y'' + amy^m(y')^3 - by(y')^2 = (by - ay^m y')^3 g\left(\dfrac{b}{a}x - \dfrac{y^m}{m}, \dfrac{y'}{by - ay^m y'}\right)$

Table 6.1. Catalogue of Second-Order ODEs

Item	d^2Y/dX^2	Simplified ODE
IIb2	$-\dfrac{(ab)^{1/2}}{4(bx^2 - ay^2)^2(ayy' - bx)^3}\{(bx^2 - ay^2)^2y'' - (bx^2 - ay^2)(xy' - y)[b - a(y')^2]$ $-2(xy' - y)(ayy' - bx)^2 + 2(ab)^{-1/2}(ayy' - bx)^3\}$	$(bx^2 - ay^2)y''$ $+(xy' - y)[a(y')^2 - b]$ $= (xy' - y)^3 g\left(bx^2 - ay^2,\right.$ $\left.\dfrac{y - xy'}{ayy' - bx}\right)$
IIb3	$\dfrac{(ab)^{1/2}\{[(bx^2 + ay^2)^2y'' - 2(xy' - y)(ayy' + bx)^2 - (bx^2 + ay^2)(xy' - y)[a(y')^2 + b]\}}{4(bx^2 + ay^2)^2(ayy' + bx)^3}$	$(bx^2 + ay^2)y''$ $-(xy' - y)[a(y')^2 + b]$ $= (xy' - y)^3 g\left(bx^2 + ay,\right.$ $\left.\dfrac{y - xy'}{ayy' + bx}\right)$

Table 6.1. Catalogue of Second-Order ODEs

Item	d^2Y/dX^2	Simplified ODE
IIc2	$\dfrac{a[[(1+x)yy'' - 2(1+x)(y')^2 + (1-a)yy']}{(1+x)^{2a-1}[ay + (1+x)y']^3}$	$(1+x)yy'' - 2(1+x)(y')^2 + (1-a)yy'$ $= \dfrac{1}{(1+x)y^2}g\left[(1+x)^a y,\ \dfrac{(1+x)yy'}{(1+x)y' + ay}\right]$
IIcs2	$-\dfrac{a[x(1+y)y'' - (1-a)x(y')^2 + 2(1+y)y']}{(1+y)^{2a-1}(1+y+axy')^3}$	$x(1+y)y'' - (1-a)x(y')^2 + 2(1+y)y'$ $= x(1+y)^2 g\left[x(1+y)^a,\ \dfrac{x(1+y)}{1+y+axy'}\right]$
IId1	$\dfrac{e^{2y/a}}{[a-(x+y)y']^3}\{-a^2y'' + [a-(x+y)y'](y')^3 + 2a(y')^2\}$	$a^2y'' - [a-(x+y)y'](y')^3 - 2a(y')^2$ $= \dfrac{1}{e^{-y/a}(y')^3}g[(x+y+a)e^{-y/a},\ Y']$
IIds1	$\dfrac{e^{2x/a}}{[ay'-(x+y)]^3}\left[a^2y'' + \dfrac{ay'-(x+y)}{y'} + 2ay'\right]$	$a^2y'' + \dfrac{ay'-(x+y)}{y'} + 2ay'$ $= e^{x/a}g[(x+y+a)e^{-x/a},\ Y']$

Table 6.1. Catalogue of Second-Order ODEs

Item	d^2Y/dX^2	Simplified ODE
IId2	$\dfrac{x^3}{[xy' - (x+y)]^3}\{x^2 y'' + [2xy' - (x+y)][xy' + (x+y)]\}$	$x^2 y'' + [2xy' - (x+y)][xy' + (x+y)]$ $= \dfrac{[xy' - (x+y)]^3}{x^3}\, g\left(\dfrac{y}{x} - \ln x, Y'\right)$
IIds2	$\dfrac{y^3}{[y-(x+y)y']^3}\{-y^2 y'' + [2y - (x+y)y'][y + (x+y)y']\}$	$y^2 y'' - [2y - (x+y)y'][y + (x+y)y']y'$ $= \dfrac{[y-(x+y)y']^3}{y^3}\, g\left(\dfrac{x}{y} - \ln y, Y'\right)$

Table 6.1. Catalogue of Second-Order ODEs

Item	d^2Y/dX^2	Simplified ODE
IIIa1	$\dfrac{BM}{(B-My')^3}[BMy'' + (BM'-B'M)(y')^2]$	$y'' + \left(\dfrac{BM'-B'M}{BM}\right)(y')^2 = \dfrac{B}{M^2}g\left(X, 1-\dfrac{M}{B}y'\right)$
IIIa2	$\dfrac{a^3 xy^{2r-1}[xyy'' - rx(y')^2 + yy']}{(by^r - axy')^3}$	$xyy'' - rx(y')^2 + yy' = \dfrac{y^{r+1}}{x}g\left(\dfrac{b}{a}\ln x - \dfrac{y^{1-r}}{1-r}, \dfrac{xy'}{y^r}\right)$
IIas2	$\dfrac{a^3 x^{2r-1}y[-xyy'' - ryy' + x(y')^2]}{(bx^r y' - ay)^3}$	$xy'' + ry' - (y')^2 = \dfrac{x(y')^2}{y}g\left(\dfrac{b}{a}\ln y - \dfrac{x^{1-r}}{1-r}, \dfrac{x^r y'}{y}\right)$
IIa3	$-\dfrac{y^{2a+1}[ax^2yy'' - a(a+1)x^2(y')^2 + a(2b+1)xyy' - b^2y^2]}{x^{2b}(axy'-by)^3}$	$x^2 y'' = yg\left(\dfrac{x^b}{y^a}, \dfrac{x}{y}y'\right)$
IIIc1	$-\dfrac{MB(MBy'' + M^2N'(y')^3 - MB'(y')^2 + M'By')}{(MNy'-B)^3}$	$MBy'' + M^2N'(y')^3 - MB'(y')^2 + M'By'$ $= \dfrac{M^2(y')^3}{B}g\left(X, \dfrac{My'}{MNy'-B}\right)$
IIIcs1	$\dfrac{MB[MBy'' - M^2N' + MB'y' - M'B(y')^2]}{(MN-By')^3}$	$MBy'' - M^2N' + MB'y' - M'B(y')^2$ $= \dfrac{M^2}{B}g\left(X, N-\dfrac{B}{M}y'\right)$

Table 6.1. Catalogue of Second-Order ODEs

Item	d^2Y/dX^2	Simplified ODE
IIIcs2	$\dfrac{a^2(m-1)x^{2m-1}y^{2r-1}[axyy'' - arx(y')^2 + amyy' - bnx^{n-m}y^{r+1}]}{(ax^m y' - bx^n y^r)^3}$	$axyy'' - arx(y')^2 + amyy' - bnx^{n-m}y^{r+1}$ $= \dfrac{y^{r+1}}{x^{2m-1}}g\left[\dfrac{bx^{n-m+1}}{a(n-m+1)} - \dfrac{y^{1-r}}{1-r},\right.$ $\left.\dfrac{y^r}{ax^m y' - bx^n y^r}\right]$
IIIcs3	$\dfrac{a^2 n y^{2r}[ax^2 y'' - ar(x^2/y)(y')^2 + a(n+1)xy' - bmy^r]}{x^n(axy' - by^r)^3}$	$ax^2 y'' - ar\dfrac{x^2}{y}(y')^2 + a(n+1)xy' - bny^r$ $= \dfrac{y^r}{x^{2n}}g\left[\dfrac{b}{a}\ln x - \dfrac{y^{1-r}}{1-r}, \dfrac{y^r}{x^n(axy' - by^r)}\right]$
IIIcs4	$\dfrac{ny^{2a+1}[ax^2yy'' - a(a+1)x^2(y')^2 + a(2b+n+1)xyy' - b(b+n)y^2]}{x^{2b+n}(axy' - by)^3}$	$ax^2yy'' - a(a+1)x^2(y')^2 + a(2b+n+1)xyy'$ $-b(b+n)y^2 = \left(\dfrac{y}{x^n}\right)^2 g\left[\dfrac{x^b}{y^a}, \dfrac{y}{x^n(axy' - by)}\right]$
IIIcs5	$\dfrac{a^2 y^{2r-1}[ax^2yy'' - arx^2(y')^2 + axyy' - bnx^n y^{r+1}]}{(bx^n y^r - axy')^3}$	$ax^2yy'' - arx^2(y')^2 + axyy' - bnx^n y^{r+1}$ $= y^{r+1}g\left(\dfrac{b}{a}\dfrac{x^n}{n} - \dfrac{y^{1-r}}{1-r}, \dfrac{y^r}{bx^n y^r - axy'}\right)$

Table 6.1. Catalogue of Second-Order ODEs

Item	d^2Y/dX^2	Simplified ODE
IIIcs6	$$\frac{a^2 y[ax^2 yy'' - ax^2(y')^2 + axyy' - bnx^n y^2]}{(bx^n y - axy')^3}$$	$$ax^2 yy'' - ax^2(y')^2 + axyy' - bnx^n y^2$$ $$= y^2 g\left(\frac{b}{a}\frac{x^n}{n} - \ln y,\; \frac{y}{bx^n y - axy'}\right)$$
IIIcs7	$$\frac{a^2(m-1)x^{m-1}y[ax^{m+1}yy'' - ax^{m+1}(y')^2 + amx^m yy' - bnx^n y^2]}{(ax^m y' - bx^n y)^3}$$	$$ax^{m+1}yy'' - ax^{m+1}(y')^2 + amx^m yy'$$ $$-bnx^n y^2 = \frac{y^2}{x^{m-1}}g\left[\frac{bx^{n-m+1}}{a(n-m+1)}\right.$$ $$\left.- \ln y,\; \frac{y}{ax^m y' - bx^n y}\right]$$
IIIcs8	$$\frac{x^{2a-n}\{-x^2 Py'' + [(2a-n-1)P + xP']xy' - a(a-n)yP - axP'y\}}{(xy' - ay)^3}$$	$$x^2 Py'' - [(2a-n-1)P + xP']xy'$$ $$+a(a-n)yP + axP'y$$ $$= X^{a-2n}P^3 g\left(\frac{y}{x^a},\; \frac{x^{a-n}P}{xy' - ay}\right)$$

Table 6.1. Catalogue of Second-Order ODEs

Item	d^2Y/dX^2	Simplified ODE
IIId1	$\dfrac{y^2[y^2y'' - x(y')^3 + y(y')^2]}{[y - (x+y)y']^3}$	$y^2y'' - x(y')^3 + y(y')^2$ $= y(y')^3 g\left[\dfrac{x}{y} - \ln y, \dfrac{yy'}{y - (x+y)y'}\right]$
IIIds1	$\dfrac{x^2(-x^2y'' - y + xy')}{[xy' - (x+y)]^3}$	$x^2y'' - xy' + y$ $= xg\left(\dfrac{y}{x} - \ln x, y' - \dfrac{x+y}{x}\right)$

Table 6.1. Catalogue of Second-Order ODEs

Item	d^2Y/dX^2	Simplified ODE
IVc4	$-\dfrac{6[a^2xyy'' - (b-ay')(bx+ay-2axy')]}{(b-ay')^3}$	$a^2xyy'' - (b-ay')(bx+ay-2axy')$ $= (b-ay')^3 g\left[bx-ay, \dfrac{x^2(b-ay')-2axy}{b-ay'}\right]$

Table 6.1. Catalogue of Second-Order ODEs

Item	d^2Y/dX^2	Simplified ODE
V1	$\dfrac{1}{E(dP/dx)^3}\left[(Q_{xx}+Q_{xy}y'+Q_yy'')\dfrac{dP}{dx}-\dfrac{E'}{E}\dfrac{dP}{dx}\left(\dfrac{dQ}{dx}\right)^2 \right.$ $\left. -(P_{xx}+P_{xy}y'+P_yy'')\dfrac{dQ}{dx}\right]$	$(Q_{xx}+Q_{xy}y'+Q_yy'')\dfrac{dP}{dx}-(P_{xx}+P_{xy}y'+P_yy'')\dfrac{dP}{dx}$ $=\dfrac{E'}{E}\dfrac{dP}{dx}\left(\dfrac{dQ}{dx}\right)^2+E\left(\dfrac{dP}{dx}\right)^3 g\left[P,\dfrac{dQ/dx}{E(dP/dx)}\right]$
V2	$\dfrac{1}{x^2a y^{2b}(bxy'+ay)^3}\{(an-bm)x^2y^2y''-b^2nx^3(y')^3$ $+[bm-an-b(2an+bm)]x^2y(y')^2$ $+[an-bm-a(an+2bm)]xy^2y'-a^2my^3\}$	$(an-bm)x^2y^2y''-b^2nx^3(y')^3$ $+[bm-an-b(2an+bm)]x^2y(y')^2$ $+[an-bm-a(an+2bm)]xy^2y'-a^2my^3$ $=(bxy'+ay)^3 g\left(x^ay^b,\dfrac{nxy'+my}{bxy'+ay}\right)$
Vs4	$\dfrac{x^3(1+y')}{(x+y)^3(xy'-y)^3}[x(x+y)y''+2(xy'-y)^2]$	$x(x+y)y''+2(xy'-y)^2=(xy'-y)^2 g\left(\dfrac{y}{x},Y'\right)$

Table 6.1. Catalogue of Second-Order ODEs

Item	d^2Y/dX^2	Simplified ODE
V5	$\left[\dfrac{x}{(x+y)(xy'-y)}\right]^3 [x(x+y)^2y'' + 2(1+y')(xy'-y)]$	$x(x+y)^2y'' + 2(1+y')(xy'-y)$ $= (1+y')^3\, g\left(\dfrac{y}{x}, \dfrac{1+y'}{xy'-y}\right)$
V6	$\left[\dfrac{x}{(x-y)(xy'-y)}\right]^3 [x(x-y)^2y'' - 2(1-y')(xy'-y)^2]$	$x(x-y)^2y'' - 2(1-y')(xy'-y)^2$ $= (1-y')^3\, g\left(\dfrac{y}{x}, \dfrac{1-y'}{xy'-y}\right)$

Tablé 7.1. Commutator and ϕ for the Four Fundamental Forms of a Two–Parameter Group

Type	e_1	e_2	$(U_1 U_2)f$	ϕ
I	0	0	0	$\neq 0$
II	0	0	0	0
III	1	0	$U_1 f$	$\neq 0$
IV	1	0	$U_1 f$	0

Table 7.2. Canonical Forms for a Second-Order ODE Invariant Under a Two-Parameter Group (Axford, 1971)

Type	Symbols	Invariants	First Differential Invariants	Second Differential Invariants	General Form Per Group	General Form for a 2-Parameter Group
I	$U_1 f = f_x$ $U_2 f = f_y$	$u_1 = y$ $u_2 = x$	$u_1^{(1)} = y'$ $u_2^{(1)} = y'$	$u_1^{(2)} = y''$ $u_2^{(2)} = y''$	$G_1(y, y', y'') = 0$ $G_2(x, y', y'') = 0$	$y'' = g(y')$
II	$U_1 f = f_y$ $U_2 f = x f_y$	$u_1 = x$ $u_2 = x$	$u_1^{(1)} = y'$ $u_2^{(1)} = y' - \dfrac{y}{x}$	$u_1^{(2)} = y''$ $u_2^{(2)} = y''$	$G_1(x, y', y'') = 0$ $G_2\left(x, y' - \dfrac{y}{x}, y''\right) = 0$	$y'' = g(x)$
III	$U_1 f = f_y$ $U_2 f = x f_x + y f_y$	$u_1 = x$ $u_2 = \dfrac{y}{x}$	$u_1^{(1)} = y'$ $u_2^{(1)} = y'$	$u_1^{(2)} = y''$ $u_2^{(2)} = x y''$	$G_1(x, y', y'') = 0$ $G_2\left(\dfrac{y}{x}, y', x y''\right) = 0$	$x y'' = g(y')$
IV	$U_1 f = f_y$ $U_2 f = y f_y$	$u_1 = x$ $u_2 = x$	$u_1^{(1)} = y'$ $u_2^{(1)} = \dfrac{y'}{y}$	$u_1^{(2)} = y''$ $u_2^{(2)} = \dfrac{y''}{y}$	$G_1(x, y', y'') = 0$ $G_2\left(x, \dfrac{y'}{y}, \dfrac{y''}{y}\right) = 0$	$y'' = y' g(x)$

Table 7.3. Canonical Coordinate
Transformations for a Two–Parameter Group

Type	Equations	
I	$\xi_1 X_x + \eta_1 X_y = 1$	$\xi_1 Y_x + \eta_1 Y_y = 0$
	$\xi_2 X_x + \eta_2 X_y = 0$	$\xi_2 Y_x + \eta_2 Y_y = 1$
II	$X = \rho(x, y)$	$\xi_1 Y_x + \eta_1 Y_y = 1$
III	$\xi_1 X_x + \eta_1 X_y = 0$	$\xi_1 Y_x + \eta_1 Y_y = 1$
	$\xi_2 X_x + \eta_2 X_y = X$	$\xi_2 Y_x + \eta_2 Y_y = Y$
IV	$\xi_1 X_x + \eta_1 X_y = 0$	$Y = \rho(x, y)$

**Table 7.4. Constraints on a Two-Parameter
Group in its Fundamental Form**

Type	Symbol Equations	ϕ Condition
I	$\xi_1\eta_{2x} + \eta_1\eta_{2y} - \xi_2\eta_{1x} - \eta_2\eta_{1y} = 0$ $\xi_1\xi_{2x} + \eta_1\xi_{2y} - \xi_2\xi_{1x} - \eta_2\xi_{1y} = 0$	$\xi_1\eta_2 - \xi_2\eta_1 \neq 0$
II	$\xi_1\eta_{2x} + \eta_1\eta_{2y} - \xi_2\eta_{1x} - \eta_2\eta_{1y} = 0$ $\xi_1\xi_{2x} + \eta_1\xi_{2y} - \xi_2\xi_{1x} - \eta_2\xi_{1y} = 0$	$\rho = \dfrac{\xi_2}{\xi_1} = \dfrac{\eta_2}{\eta_1}$
III	$\xi_1\eta_{2x} + \eta_1\eta_{2y} - \xi_2\eta_{1x} - \eta_2\eta_{1y} = \eta_1$ $\xi_1\xi_{2x} + \eta_1\xi_{2y} - \xi_2\xi_{1x} - \eta_2\xi_{1y} = \xi_1$	$\xi_1\eta_2 - \xi_2\eta_1 \neq 0$
IV	$\xi_1\eta_{2x} + \eta_1\eta_{2y} - \xi_2\eta_{1x} - \eta_2\eta_{1y} = \eta_1$ $\xi_1\xi_{2x} + \eta_1\xi_{2y} - \xi_2\xi_{1x} - \eta_2\xi_{1y} = \xi_1$	$\rho = \dfrac{\xi_2}{\xi_1} = \dfrac{\eta_2}{\eta_1}$

Table 7.5. Type, ODE, U_2f, and Canonical Variables for a Two-Parameter Group when

$$U_1f = A(x)B(y)f_y$$

Type	ODE	U_2f	Canonical Variables
I	$Y'' = g(Y')$	$\gamma(x)f_x$ $+B\left[\sigma(x) + \dfrac{\gamma A'}{A}\int \dfrac{dy}{B}\right]f_y$	$X = \dfrac{1}{A}\int \dfrac{dy}{B} - \int \dfrac{\sigma dx}{\gamma A}$ $\qquad Y = \int \dfrac{dx}{\gamma}$ $Y' = \dfrac{AB}{\gamma y' - \sigma B - (\gamma A'B/A)\int \dfrac{dy}{B}}$ $Y'' = \dfrac{\gamma AB}{\left[\gamma y' - \sigma B - (\gamma A'B/A)\int \dfrac{dy}{B}\right]^3}\left\{-\gamma ABy''\right.$ $+\gamma AB'(y')^2 + 2\gamma A'By' - \gamma' ABy'$ $+\sigma' AB^2 - \sigma A'B^2$ $\left.+\left[\gamma A'' - \dfrac{2\gamma(A')^2 B^2}{A} + \gamma' A'B^2\right]\int \dfrac{dy}{B}\right\}$

Table 7.5. Type, ODE, $U_2 f$, and Canonical Variables for a Two-Parameter Group when $U_1 f = A(x)B(y)f_y$

Type	ODE	$U_2 f$	Canonical Variables
II	$Y'' = g(X)$	$\sigma(x)Bf_y$	$X = \dfrac{\sigma}{A}$ $\qquad Y = \dfrac{1}{A}\displaystyle\int \dfrac{dy}{B}$

$$Y' = \frac{1}{B}\,\frac{Ay' - A'B\displaystyle\int \dfrac{dy}{B}}{\sigma'A - \sigma A'}$$

$$Y'' = \frac{A^3}{B^2(\sigma'A - \sigma A')^3}\left\{(\sigma'A - \sigma A')[By'' - B'(y')^2]\right.$$

$$\left. + B(\sigma A'' - \sigma''A)y' + B^2(\sigma''A' - \sigma'A'')\int \frac{dy}{B}\right\}$$

Table 7.5. Type, ODE, U_2f, and Canonical Variables for a Two-Parameter Group when $U_1f = A(x)B(y)f_y$

Type	ODE	U_2f	Canonical Variables
III	$XY'' = g(Y')$	$\gamma(x)f_x$	$X = \exp\left(\int \frac{dx}{\gamma}\right)$

$$+ B\left[\sigma(x) + \left(1 + \frac{\gamma A'}{A}\right)\int \frac{dy}{B}\right] f_y$$

$$Y = \frac{1}{A}\int \frac{dy}{B} - \exp\left(\int \frac{dx}{\gamma}\right)\int \exp\left(-\int \frac{dx}{\gamma}\right)\frac{\sigma dx}{\gamma A}$$

$$Y' = \frac{\gamma \exp\left(-\int \frac{dx}{\gamma}\right)}{AB}y' - \frac{\gamma A' \exp\left(-\int \frac{dx}{\gamma}\right)}{A^2}$$

$$- \int \exp\left(-\int \frac{dx}{\gamma}\right)\frac{\sigma dx}{\gamma A} - \frac{\sigma}{A}\exp\left(-\int \frac{dx}{\gamma}\right)$$

$$Y'' = \frac{\gamma \exp\left(-2\int \frac{dx}{\gamma}\right)}{A^3 B^2}\left\{\gamma A^2 By'' - \gamma A^2 B'(y')^2\right.$$

$$+ AB(\gamma' A - 2\gamma A' - A)y' + AB^2(\sigma A' - A\sigma')$$

$$\left. + B^2\left(\int \frac{dy}{B}\right)[-\gamma' AA' - \gamma AA'' + 2\gamma(A')^2 + AA']\right\}$$

Table 7.5. Type, ODE, U_2f, and Canonical Variables for a Two-Parameter Group when $U_1 f = A(x)B(y)f_y$

Type	ODE	U_2f	Canonical Variables
IV	$Y'' = Y'g(X)$	$B\left[\sigma(x) + \int \frac{dy}{B}\right]f_y$	$X = x$ $\qquad Y = \frac{\sigma}{A} + \frac{1}{A}\int \frac{dy}{B}$ $Y' = \frac{y'}{AB} + \frac{1}{A^2}\left(\sigma'A - \sigma A' - A'\int \frac{dy}{B}\right)$ $Y'' = \frac{y''}{AB} - \frac{B'(y')^2}{AB^2} - \frac{2A'y'}{A^2B}$ $\quad + \frac{1}{A^3}\left[A^2\sigma'' - 2\sigma'AA' - \sigma AA'' + 2\sigma(A')^2\right]$ $\quad + \frac{1}{A^3}\left[2(A')^2 - AA''\right]\int \frac{dy}{B}$

Table 7.6. Catalogue of Second-Order ODEs Invariant Under a Two-Parameter Group

Item	$U_1 f$	$U_2 f$	ODE
I1	$\eta_1(y) f_y$	$\xi_2(x) f_x$	$y'' - \dfrac{\eta_1'}{\eta_1}(y')^2 + \dfrac{\xi_2'}{\xi_2} y' = \dfrac{(y')^2}{\eta_1} g\left(\dfrac{\xi_2 y'}{\eta_1}\right)$
I2	f_y	$x f_x$	$x^2 y'' = g(x y')$
I2s	f_x	$y f_y$	$y'' = y' g\left(\dfrac{y'}{y}\right)$
I3	$x f_y$	$[\sigma(x) + y] f_y$	$x y'' = \sigma' + g\left(\dfrac{x y' - y - \sigma}{x}\right)$
I3s	$y f_x$	$[\sigma(y) + x] f_x$	$y y'' = -\sigma'(y')^3 + (y')^3 g\left[\dfrac{y - (x + \sigma) y'}{y y'}\right]$

Table 7.6. Catalogue of Second-Order ODEs Invariant Under a Two-Parameter Group

Item	$U_1 f$	$U_2 f$	ODE
I4	$y f_y$	$x f_x + y\sigma(x) f_y$	$y'' - \dfrac{(y')^2}{y} + \dfrac{y' - y\sigma'}{x} = \dfrac{y}{x^2} g\left(\dfrac{xy' - y\sigma}{y}\right)$
I4s	$x f_x$	$x\sigma(y) f_x + y f_y$	$y'' + \dfrac{y'}{x} - \dfrac{(y')^2}{y} + \dfrac{x\sigma'}{y}(y')^3 = \dfrac{x(y')^3}{y^2} g\left(\dfrac{y - x\sigma y'}{xy'}\right)$
I5	$y f_y$	$x f_x$	$x^2 y'' = yg\left(\dfrac{xy'}{y}\right)$
I6	$x f_x + y f_y$	$(x+y) f_x + (y-x) f_y$	$(x^2 + y^2)^2 y'' = (y - xy')^3 g\left(\dfrac{y - xy'}{x + yy'}\right)$

Table 7.6. Catalogue of Second-Order ODEs Invariant Under a Two-Parameter Group

Item	$U_1 f$	$U_2 f$	ODE
II1	$\eta_1(x)f_y$	$\eta_2(x)f_y$	$(\eta_1\eta_2' - \eta_2\eta_1')y'' + (\eta_2\eta_1'' - \eta_1\eta_2'')y' + (\eta_1'\eta_2'' - \eta_2'\eta_1'')y$ $= \left(\dfrac{\eta_1\eta_2' - \eta_2\eta_1'}{\eta_1}\right)^3 g\left(\dfrac{\eta_2}{\eta_1}\right)$
II2	yf_y	$y\sigma(x)f_y$	$yy'' - (y')^2 - \dfrac{\sigma''}{\sigma'}yy' = y^2 g(x)$
II2s	xf_x	$x\sigma(y)f_x$	$xy'' + y' + \dfrac{\sigma''}{\sigma'}x(y')^2 = x^2(y')^3 g(y)$
II3	xf_y	$x^2 f_y$	$x^2 y'' - 2xy' + 2y = g(x)$
II3s	yf_x	$y^2 f_x$	$y^2 y'' + 2(y')^2 - 2x(y')^3 = (y')^3 g(y)$

Table 7.6. Catalogue of Second-Order ODEs Invariant Under a Two-Parameter Group

Item	$U_1 f$	$U_2 f$	ODE
II4	xyf_y	yf_y	$yy'' - (y')^2 = y^2 g(x)$
II4s	xyf_x	xf_x	$xy'' + y' = x^2(y')^3 g(y)$
II5	$\dfrac{1}{y}f_y$	$\dfrac{x}{y}f_y$	$yy'' + (y')^2 = g(x)$
II5s	$\dfrac{1}{x}f_x$	$\dfrac{y}{x}f_x$	$xy'' - y' = (y')^3 g(y)$
II6	$xy^2 f_y$	$x^2 y^2 f_y$	$x^2 yy'' - 2x^2(y')^2 - 2xyy' - 2y^2$ $= y^3 g(x)$
II6s	$x^2 y f_x$	$x^2 y^2 f_x$	$xy^2 y'' + 2y^2 y' + 2xy(y')^2 + 2x^2(y')^3$ $= x^3(y')^3 g(y)$
II7	$(x^2 + xy)f_x + (xy + y^2)f_y$	$(x^2 - xy)f_x + (xy - y^2)f_y$	$x^3 y'' = (xy' - y)^3 g\left(\dfrac{y}{x}\right)$

Table 7.6. Catalogue of Second-Order ODEs Invariant Under a Two-Parameter Group

Item	U_1f	U_2f	ODE
III1	f_y	$\gamma(x)f_x + yf_y$	$\gamma y'' + (\gamma' - 1)y' = y'g\left[\gamma y' \exp\left(-\int \frac{dx}{\gamma}\right)\right]$
III1s	f_x	$xf_x + \gamma(y)f_y$	$\gamma y'' + (1-\gamma')(y')^2 = (y')^2 g\left[\frac{y'}{\gamma}\exp\left(\int \frac{dy}{\gamma}\right)\right]$
III2	xf_y	$\gamma(x)f_x + \left(1+\frac{\gamma}{x}\right)yf_y$	$\gamma^2 y'' + \frac{\gamma}{x^2}(x\gamma' - 2\gamma - x)(xy' - y)$ $= x\exp\left(\int \frac{dx}{\gamma}\right) g\left[\frac{\gamma}{x^2}(xy' - y)\exp\left(-\int \frac{dx}{\gamma}\right)\right]$
III2s	yf_x	$\left(1+\frac{\gamma}{y}\right)xf_x + \gamma(y)f_y$	$\gamma^2 y'' + \frac{\gamma}{y^2}(y\gamma' - 2\gamma - y)(xy' - y)(y')^2$ $= y(y')^3 \exp\left(\int \frac{dy}{\gamma}\right) g\left[\frac{\gamma(xy' - y)}{y^2 y'}\exp\left(-\int \frac{dy}{\gamma}\right)\right]$

Table 7.6. Catalogue of Second-Order ODEs Invariant Under a Two-Parameter Group

Item	$U_1 f$	$U_2 f$	ODE
III3	f_y	$xf_x + yf_y$	$xy'' = g(y')$
III3s	f_x	$xf_x + yf_y$	$yy'' = g(y')$
III4	xf_y	$-xf_x$	$x^2 y'' = g(xy' - y)$
III4s	yf_x	$-yf_y$	$y^2 y'' = (y')^3 g\left(\dfrac{xy' - y}{y'}\right)$
III5	f_y	$x^2 f_x + y(1+x)f_y$	$x^3 y'' = (xy' - y)g[(xy' - y)\exp(1/x)]$
III5s	f_x	$x(1+y)f_x + y^2 f_y$	$y^3 y'' = (xy' - y)(y')^2 g\left[\dfrac{xy' - y}{y'}\exp(1/y)\right]$

Table 7.6. Catalogue of Second-Order ODEs Invariant Under a Two-Parameter Group

Item	U_1f	U_2f	ODE
IV1	$B(y)f_y$	$B\left(\int \dfrac{dy}{B}\right)f_y$	$y'' - \dfrac{B'}{B}(y')^2 = y'g(x)$
IV1s	$B(x)f_x$	$B\left(\int \dfrac{dx}{B}\right)f_x$	$y'' + \dfrac{B'}{B}y' = (y')^2 g(y)$
IV2	$A(x)f_y$	yf_y	$y'' - \dfrac{2A'}{A}y' + \left[\dfrac{2(A')^2 - AA''}{A^2}\right]y = (Ay' - A'y)g(x)$
IV2s	$A(y)f_x$	xf_x	$y'' + \dfrac{2A'}{A}(y')^2 - \left[\dfrac{2(A')^2 - AA''}{A^2}\right]x(y')^3$ $$= (A - A'xy')(y')^2 g(y)$$
IV3	f_x	xf_x	$y'' = (y')^2 g(y)$
IV3s	f_y	yf_y	$y'' = y'g(x)$

Table 7.6. Catalogue of Second-Order ODEs Invariant Under a Two-Parameter Group

Item	$U_1 f$	$U_2 f$	ODE
IV4	$f_x + f_y$	$xf_x + xf_y$	$y'' = (y'-1)^2 g(y-x)$
IV4s	$f_x + f_y$	$yf_x + yf_y$	$y'' = (y'-1)^2 y' g(y-x)$
IV5	$f_x + yf_y$	$xf_x + xyf_y$	$yy'' - (y')^2 = (y'-y)^2 g(x - \ln y)$
IV5s	$xf_x + f_y$	$xyf_x + yf_y$	$xy'' + y' = (xy'-1)^2 y' g(y - \ln x)$
IV6	$f_x + xf_y$	$xf_x + x^2 f_y$	$y'' - 1 = (y'-x)^2 g(2y - x^2)$
IV6s	$yf_x + f_y$	$y^2 f_x + yf_y$	$y'' + (y')^3 = (yy'-1)^2 y' g(2x - y^2)$
IV7	$(x+y)f_x$	xf_x	$y'' = (y')^2 (xy'-y)g(y)$
IV7s	$(x+y)f_y$	yf_y	$y'' = (xy'-y)g(x)$

Appendix E

Answers to Selected Problems

2.3. (1) $Uf = \dfrac{1}{2}yf_x - \dfrac{y^2}{2x}f_y$

(2) $Uf = f_x - \dfrac{1+y}{x}f_y$

(3) $Uf = -x^2f_x - xyf_y$

2.4. (1) $\delta x = 2x\delta\alpha,\ \delta y = 3y\delta\alpha,\ Uf = 2xf_x + 3yf_y$

2.5. (1) $x_1 = x + \dfrac{1}{2}\alpha y - \dfrac{1}{8}\alpha^2\dfrac{y^2}{x} + \cdots$

$\qquad\qquad y_1 = y - \dfrac{1}{2}\alpha\dfrac{y^2}{x} + \dfrac{3}{8}\alpha^2\dfrac{y^3}{x^2} + \cdots$

2.8. For (1) with $\alpha = e^{\tilde{\alpha}}$

$\qquad x_1 = x + 2\tilde{\alpha}x + 2\tilde{\alpha}^2x + \cdots$

$\qquad y_1 = y + 3\tilde{\alpha}y + \dfrac{9}{2}\tilde{\alpha}^2y + \cdots$

$\qquad v = \dfrac{1}{3}\ln y,\ u = \dfrac{y}{x^{3/2}}$

$\qquad u_1 = u,\ v_1 = v + \tilde{\alpha}$

2.9. (3) $x_1 = e^\alpha(x\cos\alpha - y\sin\alpha)$

$$y_1 = e^\alpha (x \sin \alpha + y \cos \alpha)$$

3.2. $G(y + 2 \ln x, y' e^{-y/2}, y'' e^{-y/2}) = 0$

3.3. $G(y + 2 \ln(x \ln x), 2 \ln x + x(\ln x)y', x(\ln x)^2(xy'' + y'), f) = 0$

3.7. $G[x^2 + y^2, x(1 - f^2)^{1/2} + yf] = 0$

3.8. $V(x, t) = V_0(x - V^n t)$

4.2. (1) $\eta' = y', \quad \eta'' = -y''$

(2) $\eta' = -y' - (y')^2, \quad \eta'' = -(2 + 3y')y''$

4.4. $\Omega(f) = -f$

4.5. $Uf = mx f_x + ny f_y$

$\eta' = (n - m)y', \quad \eta'' = (n - 2m)y''$

5.1. (1) $G\left(xy, \dfrac{x}{y}\dfrac{xy' - y}{xy' + y}\right) = 0$

(2) $G[x(1 + y), xy' + y] = 0$

(3) $G\left(\dfrac{y}{x}, xy' - y\right) = 0$

5.3. (1) $G(xy, x^2 y') = 0$

(2) $G\left[y, \dfrac{(x + y)y'}{y' + 1}\right] = 0$

(3) $G\left[\dfrac{\exp[\tan^{-1}(y/x)]}{(x^2 + y^2)^{1/2}}, \dfrac{xy' - y}{yy' + x}\right] = 0$

5.7. (1) $X = \dfrac{1}{2} \ln x, \quad Y = \dfrac{y}{x^{3/2}}, \quad Y' = 2Y^{1/3} g(Y) - 3Y$

(2) $X = \ln(x + y), \quad Y = y, \quad Y' = g(Y)$

5.9. (3) $\dfrac{x}{y} = c - 3 \int \exp[(\sin^{-1} Y)^2] \dfrac{dY}{Y}$ where $Y = (xy)^{1/3}$

5.10. (2) $\dfrac{1}{x} = c - a^{-1/2} \displaystyle\int \dfrac{X^{1/2}dX}{(-\ln X)^{1/2}}$ where $X = (y/x)$

5.13. $x^2 - 2ay^2 + 4a^2 = 0$

5.17. $e^{-my} = cx^{am} - \dfrac{mb}{e - am + 1}x^{e+1}$

5.19. $e^x = ce^{y/x} - \left(\dfrac{y}{x}\right)^2 - 2\dfrac{y}{x} - 2$

5.21. $\ln x = e^{-y^4/4}\left(c + \displaystyle\int e^{X^4/4}dX\right)$ where $X = y$

6.1. $G\left(\dfrac{y}{x}, xy' - y, x^3y''\right) = 0$

6.3. $y = \left(\dfrac{3}{2}\dfrac{x^3}{1-bx}\right)^{1/2}$

6.5. $G\left[y^2 - x^2, (x+y)^2\dfrac{1-y'}{1+y'}, \dfrac{y''}{[1-(y')^2]^{3/2}}\right] = 0$

6.6. $G_1\left[x^2 - c^2t^2, (x+ct)^2\left(\dfrac{x'-c}{x'+c}\right)\right] = 0$

$G_2\left[x^2 - c^2t^2, (x+ct)^2\left(\dfrac{x'-c}{x'+c}\right), \dfrac{x''}{[c^2-(x')^2]^{3/2}}\right] = 0$ where $x' = (dx/dt)$,

$x'' = (d^2x/dt^2)$

6.9. (5) $y = x\tan\left\{b - \dfrac{1}{2}\sin^{-1}[2a(x^2 + y^2) - 1]\right\}$

(6) $y = b + \ln(ax + x^2)$

6.12. $xy^2 = be^x + a - \dfrac{2}{3}(x^3 + 3x^2 + 6x)$

6.17. $x = y + c_1\sin(y - c_2)$

6.20. $y_1 = y, \quad y_2 = y', \quad y_3 = y''$

$\xi f_{1,x} + \eta_1 f_{1,y_1} + \eta_2 f_{1,y_2} + \eta_3 f_{1,y_3} + \eta_3' f_{1,y_3'} = 0$

$$\eta_2 = \eta_1', \quad \eta_3 = \eta_2'$$

6.22. $-a_1(a_2''' + a_3'') + a_1'(a_2'' - 2a_3') + 3a_1''(-a_2 + 2a_1') + 3a_1'''a_2 = 0$

$-3a_1^2 a_4'' + a_1(2a_2 a_3'' - a_2 a_2''' - 2a_3 a_3' + 4a_2' a_3' + a_2'' a_3 - 2a_2' a_2'') - 2a_1' a_2 a_3' +$

$a_1' a_2 a_2'' + 3a_1 a_1'' a_4 = 0$

7.1. $y = [b \exp(a^{-1} e^x)]$

7.4. $x = b + a \int \exp(-e^{-X}) dX$ where $X = x - \ln y$, or $x = b + a E_1(y e^{-x})$,
where E_n is the exponential integral

7.6. Type II

$$x^3 y'' = (xy' - y)^3 g\left(\frac{y}{x}\right)$$

$$\rho = \frac{x - y}{x + y} = X$$

$$Y = -\frac{1}{x + y}$$

7.7. Type III

$$X = \frac{y}{x(x + y)}, \quad Y = -\frac{1}{x}$$

$$xy'' = \frac{2(1 + y')(xy' - y)}{x + y} + \frac{(x + y)^3}{x^2 y} g\left[\frac{x^2 y' - y^2 - 2xy}{(x + y)^2}\right]$$

7.11. Type II

$$X = \rho = (x^2 + y^2)^{1/2}, \quad Y = -\tan^{-1}\left(\frac{y}{x}\right)$$

$$\frac{(x^2 + y^2)^2}{(x + yy')^3} y'' + 2\frac{y - xy'}{x + yy'} + \left(\frac{y - xy'}{x + yy'}\right)^3 = g(x^2 + y^2)$$

7.15. Type I

$$xy \left(\frac{y}{x}\right)^n (x^2+y^2)^2 y'' + (xy'-y)^2 [x-y+(x+y)y'] \left[n(x^2+y^2) - \left(\frac{y}{x}\right)^n (x^2-y^2)\right]$$

$$-xy \left(\frac{y}{x}\right)^n (x^2+y^2)(xy'-y)[1+(y')^2]$$

$$= \frac{(xy)^2 [x-y+(x+y)y']^3}{x^2+y^2} g \left[\frac{x^2+y^2}{xy} \left(\frac{y}{x}\right)^n \frac{xy'-y}{x-y+(x+y)y'}\right]$$

Index